Fundamentals of Agriculture

About the Editors

Dr. Sapna hails from Rewari, a district in Haryana. She is currently employed as a Senior Research Fellow (DUS) in the Crop Improvement Division at ICAR-IIWBR, Karnal. She graduated (B.Sc. biotechnology) from MDU Rohtak, Haryana. She earned M.Sc. and Ph.D in plant physiology from HAU in Hisar, Haryana. She qualified ICAR (ASRB)-NET in Plant Physiology, CISR Net in Life Sciences & GATE in Botany and Biotechnology. In addition to 12 articles, she has authored 14 research and review papers, 5 book chapters, and 15 abstracts. She has attended numerous conferences seminars, and congresses both nationally and internationally. She has been awarded Acedemic excellance award, best Article Award, Innovative Article Award & Young Scientist Award in 6th International Conference on Histolic Innovations and Technological Advance for Sustainable Agriculture.

Vijay Kumar was born and brought up in Muzaffarnagar district, located in the state of Uttar Pradesh. He is currently stationed in ICAR-Sugarcane Breeding Institute, Regional Centre, Karnal serving as a Technical Officer in the Plant Breeding Division. He obtained a Bachelor of Science degree in Agriculture from SVBP University of Agriculture & Technology, Modipuram, Meerut, Uttar Pradesh. He also accomplished a Master of Science degree in Genetics & Plant Breeding at SHUATS, Naini Prayagraj, Uttar Pradesh. He has authored 2 articles, 6 research papers, and 3 abstracts. He has participated in numerous national and international seminars congresses, and conferences. He received the Young Scientist Award at the 6th International Conference on Histolic Innovations and Technological Advance for Sustainable Agriculture.

Dr. Vijay Kumar Vimal has completed his B.Sc. (Agriculture) from P. G. College Ghazipur (U.P.). He has completed M.Sc.(Ag) and Ph.D. (Vegetable Science) Vegetable Science from ANDUAT Kumarganj Ayodhya (U.P.). He has qualified ASRB NET in Vegetable Science. He Published 5 Book, 5 Manual, 10 Research papers, 25 Popular articles, 10 Abstracts and participated 25 Seminar/Symposia/Conference. Currently he is working as Scientist (Horticulture) at Krishi Vigyan Kendra Kotwa, Azamagrh-I which is governed by Acharya Narendra Deva University of Agriculture and Technology, Kumarganj, Ayodhya, (U.P.) India.

Aastha Dubey, a passionate scholar, completed her graduation in Agriculture from Sharda University, Greater Noida, U.P. where she was honored as a Gold Medalist. Continuing her academic journey, she pursued her Masters in Fruit Science (Horticulture) from Sardar Vallabhbhai Patel University of Agriculture and Technology, Meerut, U.P. Her dedication in research is evident through her contributions to numerous book chapters, review articles and research papers as well as securing a patent. She has actively participated in various national and international conferences. She has participated and won second prize (zonal level) at a Student research convention hosted by AIU (Association of Indian Universities). She is currently pursuing Ph.D. (Horticulture) Fruit Science from Galgotias University, Greater Noida, U.P.

Prasun Kumar Singh has completed his B.Sc. (Hons.) Agriculture and M.Sc. (Agriculture) Agronomy from Sharda University, Greater Noida, Uttar Pradesh. Currently, he is pursuing Ph.D. in Agronomy from Galgotias University, Greater Noida, Uttar Pradesh. He has contributed in various Articles, Abstracts and book chapters. He has actively participated in various national and international conferences and won second prize (zonal level) at a Student Research Convention hosted by Association of Indian Universities.

1168, Sector 13, Urban Estate
Karnal-132 001, Haryana
Tel: 91-84470 75807, 18440 41168
Email: contentvibesppa@gmail.com
www.contentvibes.in

Print ISBN: 978-93-6134-590-6

ebook ISBN: 978-93-6134-6-58-3

Fundamentals of Agriculture

Sapna
Vijay Kumar
Vijay Kumar Vimal
Aastha Dubey
Prasun Kumar Singh

Preface

Agriculture, the backbone of human civilization, has evolved tremendously from its humble beginnings to the sophisticated and diverse practice it is today. The fundamental principles of agriculture encompass a vast array of disciplines, integrating science, technology, and traditional knowledge to ensure the sustainability and productivity of agricultural systems. This book, "Fundamentals of Agriculture," aims to provide a comprehensive overview of the essential concepts and practices that form the foundation of modern agriculture.

Purpose and Scope

The purpose of this book is to serve as an introductory text for students, educators, and practitioners in the field of agriculture. It covers a broad spectrum of topics, including soil science, crop production, animal husbandry, agricultural economics, and sustainable practices. By presenting these subjects in a coherent and accessible manner, the book seeks to equip readers with the knowledge necessary to understand and engage in agricultural activities effectively.

Structure of the Book

"Fundamentals of Agriculture" is organized into several sections, each dedicated to a specific aspect of agriculture. The structure is designed to guide readers from basic concepts to more complex topics, ensuring a gradual and comprehensive learning experience:

1. **Introduction to Agriculture:** This section provides an overview of the history and development of agriculture, highlighting its significance in human society and the various types of agricultural systems practiced around the world.
2. **Soil Science and Management:** Here, readers will explore the properties of soil, soil fertility, and the techniques used for soil conservation and management. Understanding soil health is crucial for successful crop production.
3. **Crop Production:** This section delves into the principles of crop cultivation, including the selection of crops, planting methods, irrigation,

pest management, and harvesting. It emphasizes the importance of optimizing crop yields while maintaining environmental sustainability.

4. **Animal Husbandry:** Focusing on the care and management of livestock, this section covers breeding, nutrition, disease control, and the economic aspects of animal production. The role of livestock in agricultural systems and food security is also discussed.
5. **Agricultural Economics and Policy:** Readers will learn about the economic principles that underpin agricultural practices, market dynamics, and the impact of agricultural policies. This section also addresses global trade, subsidies, and the economic challenges faced by farmers.
6. **Sustainable Agriculture:** Emphasizing the need for sustainable practices, this section covers organic farming, agroecology, and the integration of modern technology with traditional knowledge to promote environmental health and resource conservation.

The creation of this book would not have been possible without the contributions of numerous experts in the field of agriculture. We extend our heartfelt gratitude to the researchers, educators, and practitioners who provided valuable insights and feedback. Special thanks are due to the institutions and organizations that supported this endeavour, offering resources and expertise essential for the completion of this work. As we face the challenges of a growing global population, climate change, and resource depletion, the role of agriculture has never been more critical. This book aims to inspire a deeper understanding and appreciation of agriculture's fundamental principles, encouraging innovation and sustainable practices. It is our hope that "Fundamentals of Agriculture" will serve as a valuable resource for anyone involved in the vital task of feeding the world and preserving our planet for future generations.

Sapna
Vijay Kumar
Vijay Kumar Vimal
Aastha Dubey
Prasun Kumar Singh

Contents

1

Principles of Crop Production and Management

D.B. Patoliya

Agriculture Assistant, Department of Horticulture, BACA, Anand Agricultural University, Anand, Gujarat

Abstract

The principles of crop production and management constitute the cornerstone of modern agriculture, embodying a delicate balance between tradition and innovation. Rooted in centuries of agricultural wisdom and scientific inquiry, these principles guide farmers in harnessing the earth's resources to nurture crops and sustain life. From soil preparation to post-harvest handling, each stage of the crop production cycle reflects a commitment to efficiency, sustainability, and resilience. By understanding and adhering to these principles, farmers can optimize yields, minimize environmental impacts, and ensure food security for present and future generations. In essence, the abstract of the principles of crop production and management encapsulates a holistic vision of agriculture, where stewardship of the land converges with the imperatives of productivity and sustainability.

Keywords: *Production, management, efficiency, principle*

Crop Production

Crop production refers to the cultivation and management of plants for the purpose of generating agricultural yields. It encompasses a wide range of activities, starting from land preparation and seed selection to harvesting and post-harvest handling. Crop production involves the application of scientific principles, traditional knowledge, and modern technologies to optimize plant growth, minimize losses, and maximize yields. The goal of crop production is not only to meet the demands for food, fibre, and fuel but also to do so in a sustainable manner that preserves the health of the soil, water resources, and surrounding ecosystems. It is a fundamental component of agriculture, playing a pivotal role in sustaining human life and supporting economic development around the globe.

Principles of Crop Production and Management

The principles of crop production encapsulate the fundamental guidelines and practices that govern the cultivation and management of crops. These principles serve as a framework for farmers to optimize agricultural productivity, sustainability, and profitability. Here are some key principles:

1. **Soil Health**: Healthy soil forms the foundation of successful crop production. Principles related to soil health emphasize the importance of maintaining soil fertility, structure, and biological activity. Practices such as soil testing, organic matter incorporation, and conservation tillage are employed to enhance soil health and productivity. Soil health is a foundational principle in the realm of crop production, influencing every aspect of plant growth and agricultural sustainability. Healthy soil serves as a reservoir of essential nutrients, a habitat for beneficial microorganisms, and a medium for root development. The principles of soil health emphasize the importance of maintaining soil fertility, structure, and biological activity through practices such as organic matter incorporation, conservation tillage, and cover cropping. By replenishing organic matter, optimizing soil structure, and fostering a diverse microbial community, farmers can enhance soil health and productivity. Soil testing is another key principle, enabling farmers to assess nutrient levels, pH balance, and other soil properties, guiding the application of fertilizers and soil amendments. Moreover, conservation practices such as crop rotation, agroforestry, and erosion control help protect soil from degradation, preserving its long-term productivity and resilience. Embracing these principles of soil health not only promotes higher crop yields but also mitigates environmental impacts, enhances water retention, and fosters ecosystem biodiversity, laying the groundwork for sustainable agriculture now and in the future.

2. **Seed Selection and Management**: Choosing high-quality seeds adapted to local growing conditions is essential for achieving optimal yields. Principles of seed selection focus on factors such as seed variety, genetic traits, germination potential, and seed treatment to ensure uniform stand establishment and crop vigour. Seed selection and management constitute vital principles within the domain of crop production, profoundly influencing the success and sustainability of agricultural

endeavours. The principle underscores the critical importance of choosing seeds that are well-adapted to local climatic conditions, soil types, and farming practices. Selecting high-quality seeds with desirable traits such as disease resistance, stress tolerance, and high yield potential is fundamental to achieving optimal crop performance. Moreover, seed management encompasses various practices aimed at maximizing seed viability, germination rates, and uniform stand establishment. These practices may include seed treatment, storage, and proper handling techniques to ensure the preservation of seed quality from planting to emergence. By adhering to the principles of seed selection and management, farmers can mitigate risks associated with poor seed quality, enhance crop resilience to biotic and abiotic stresses, and ultimately, optimize yields while promoting agricultural sustainability. Through continuous research, innovation, and informed decision-making, seed selection and management remain integral pillars in the pursuit of agricultural productivity and food security worldwide.

3. **Nutrient Management**: Proper nutrition is critical for plant growth and development. Principles of nutrient management involve assessing soil nutrient levels, understanding crop nutrient requirements, and applying fertilizers or soil amendments judiciously to optimize nutrient availability and uptake. Nutrient management stands as a cornerstone principle within the framework of crop production, governing the optimal provision of essential elements necessary for plant growth, development, and productivity. This principle emphasizes the judicious assessment of soil nutrient levels, crop nutrient requirements, and the application of fertilizers or soil amendments to maintain soil fertility and maximize crop yields. By understanding the dynamic interplay between soil properties, nutrient availability, and plant demand, farmers can tailor nutrient management practices to suit specific crop needs and environmental conditions. Furthermore, nutrient management extends beyond mere supplementation, encompassing strategies to enhance nutrient use efficiency, minimize nutrient losses, and mitigate environmental impacts such as nutrient runoff and leaching. Adopting precision agriculture techniques, such as variable rate fertilization and site-specific nutrient application, enables farmers to optimize resource utilization and minimize input costs while promoting sustainable agricultural practices. Ultimately, nutrient management plays a pivotal role in ensuring the long-term productivity, profitability, and environmental stewardship of agricultural systems, underscoring its significance as a fundamental principle in crop production.

4. **Water Management**: Efficient water management is essential for maximizing crop yields while conserving water resources. Principles of water management include irrigation scheduling, water-saving techniques such as drip irrigation and mulching, and optimizing soil water-holding capacity to minimize water stress and maximize water use efficiency. Water management serves as a crucial principle in the realm of crop production, encompassing strategies to optimize water use efficiency, mitigate water stress, and sustainably utilize water resources. This principle recognizes the paramount importance of water in supporting plant growth, nutrient uptake, and overall crop productivity. By implementing efficient irrigation systems, such as drip irrigation or precision sprinklers, farmers can deliver water directly to the root zone, minimizing losses to evaporation and runoff while maximizing irrigation efficiency. Additionally, employing water-saving techniques like mulching, rainwater harvesting, and soil moisture monitoring allows farmers to conserve water and adapt to fluctuating precipitation patterns. Timely irrigation scheduling based on crop water requirements and soil moisture levels further enhances water management practices, ensuring that crops receive adequate moisture without excess or deficit. Moreover, integrating water conservation practices with soil and nutrient management strategies promotes synergistic benefits, fostering resilient agricultural systems capable of withstanding droughts and water scarcity. By embracing the principles of water management, farmers can enhance crop yields, optimize resource utilization, and contribute to the sustainability of agricultural production in the face of climate variability and water scarcity challenges.

5. **Pest, Disease, and Weed Management**: Effective pest, disease, and weed management strategies are crucial for protecting crop health and minimizing yield losses. Principles of integrated pest management (IPM) emphasize a combination of cultural, biological, and chemical control methods to manage pests, diseases, and weeds while minimizing environmental impacts. Pest, disease, and weed management are integral principles in crop production, aimed at safeguarding crop health and maximizing yields while minimizing environmental impacts. These principles recognize the pervasive threats posed by pests, diseases, and weeds, which can significantly reduce crop productivity if left unmanaged. Integrated pest management (IPM) serves as a cornerstone approach, emphasizing the use of a combination of cultural, biological, and chemical control methods to mitigate pest and disease pressures effectively. Cultural practices such as crop rotation, sanitation, and

planting resistant varieties help disrupt pest and disease cycles and reduce reliance on chemical pesticides. Biological control agents, such as beneficial insects and microorganisms, offer natural solutions for pest suppression while minimizing harm to non-target organisms and ecosystems.

Similarly, weed management principles focus on preventing weed establishment and proliferation through cultural practices like crop rotation, mulching, and mechanical weed control. Additionally, the integration of herbicides with diverse weed management tactics helps manage weed populations while minimizing herbicide resistance and environmental contamination. Timely weed monitoring and management interventions are essential for maintaining crop competitiveness, optimizing resource utilization, and preserving soil health. By embracing principles of integrated pest, disease, and weed management, farmers can effectively mitigate crop losses, reduce pesticide inputs, and foster resilient agroecosystems. Moreover, promoting biodiversity, enhancing soil health, and minimizing environmental impacts contribute to the sustainability and long-term viability of agricultural production systems. Through continuous monitoring, innovation, and adaptive management, farmers can navigate pest, disease, and weed challenges while promoting agricultural productivity, environmental stewardship, and food security.

6. **Crop Monitoring and Maintenance**: Regular monitoring of crops enables farmers to assess growth dynamics, detect potential issues, and implement timely interventions. Principles of crop monitoring involve utilizing tools such as remote sensing, drones, and sensor-based technologies to monitor crop health, nutrient status, and pest incidence. Additionally, timely maintenance activities such as pruning, thinning, and trellising promote optimal crop growth and development. Crop monitoring and maintenance represent indispensable principles in the realm of crop production, facilitating the proactive management of crops to optimize growth, health, and yield potential. These principles underscore the importance of regular surveillance and intervention throughout the crop growth cycle to detect and address potential issues promptly. Various monitoring techniques, including remote sensing, drones, and sensor-based technologies, provide farmers with real-time insights into crop health, nutrient status, and environmental conditions. By utilizing these tools, farmers can identify anomalies such as nutrient deficiencies, pest infestations, or water stress early on, enabling timely interventions to mitigate risks and minimize yield losses. Maintenance

activities such as pruning, thinning, and trellising play a crucial role in optimizing crop development and productivity. Pruning encourages proper canopy structure, light penetration, and airflow, promoting photosynthesis, fruit set, and overall plant vigour. Thinning facilitates adequate spacing between plants, reducing competition for resources and enhancing individual plant growth and yield potential. Trellising supports vine crops, such as tomatoes and cucumbers, improving fruit quality, facilitating harvesting, and minimizing disease incidence by keeping fruits off the ground. Furthermore, precision agriculture technologies enable site-specific management practices tailored to the unique characteristics of each field or crop zone. By integrating crop monitoring data with agronomic knowledge and decision support systems, farmers can optimize resource allocation, minimize input costs, and maximize returns on investment. Overall, crop monitoring and maintenance principles empower farmers to proactively manage their crops, optimize production efficiency, and adapt to dynamic environmental conditions, ultimately contributing to sustainable agricultural practices and food security.

7. **Harvesting and Post-Harvest Management**: Proper harvesting and post-harvest handling practices are essential for preserving crop quality and minimizing losses. Principles of harvesting and post-harvest management include harvesting at optimal maturity, careful handling to prevent damage, and employing techniques such as cleaning grading, and storage to maintain product integrity and prolong shelf life. Harvesting and post-harvest management represent critical principles within the overarching framework of crop production, ensuring the preservation of crop quality and maximizing yield value. Harvesting marks the culmination of the growing season, requiring careful planning and execution to optimize crop maturity, minimize losses, and maintain product integrity. Timing is crucial during harvesting, as crops should be harvested at peak maturity to achieve optimal yield and quality while avoiding over ripening or deterioration. Depending on the crop type and harvesting method, specialized equipment such as combines, harvesters or hand tools may be employed to facilitate efficient harvesting operations.

   Following harvesting, post-harvest management practices come into play to preserve crop quality and extend shelf life. These practices encompass a range of activities, including cleaning, sorting, grading, and packaging, aimed at removing debris, separating defective or damaged produce, and preparing crops for market distribution. Proper handling techniques are crucial during post-harvest operations to minimize physical damage and reduce susceptibility to decay or spoilage. Cold storage facilities, controlled atmosphere storage, and refrigerated transport systems help maintain product freshness and quality, particularly for perishable crops such as fruits and vegetables. Moreover, post-harvest management extends beyond immediate handling to include value-added processing, such as drying, milling, or canning, to enhance product marketability and shelf stability. Quality assurance measures, including traceability systems and food safety protocols, are integral components of post-harvest management, ensuring compliance with regulatory standards and consumer expectations. By implementing effective harvesting and post-harvest management practices, farmers can maximize the value of their crops, minimize losses, and meet consumer demand for high-quality, safe food products, thereby contributing to the overall sustainability and profitability of agricultural production systems.

Conclusion

In conclusion, the principles of crop production constitute a comprehensive framework that guides farmers in cultivating crops sustainably, optimizing yields, and ensuring food security. From soil preparation to post-harvest management, each principle plays a vital role in the success and resilience of agricultural systems. By prioritizing soil health, water efficiency, nutrient management, and pest control, farmers can mitigate environmental degradation, minimize resource wastage, and promote long-term sustainability. Embracing technological advancements and innovative practices empowers farmers to adapt to changing climatic conditions and meet the evolving demands of a growing population. Furthermore, the principles of crop production underscore the interconnectedness of agricultural practices with broader ecological and socioeconomic contexts, highlighting the importance of holistic approaches that balance productivity with environmental stewardship and social responsibility. Through continuous learning, adaptation, and collaboration, stakeholders across the agricultural value chain can work together to build resilient and sustainable food systems capable of nourishing present and future generations.

References

Arnon, I. (1971). Crop production In dry regions. Volume 1: Background and principles. London, Leonard Hill..

Brown, T., & Haward, R. (2001). Principles and practice in organic crop production. Pesticide Outlook, 12(2), 64-67.

Cardwell, V. B. (1985). Principles approach in teaching crop management courses. Journal of Agronomic Education, 14(2), 98-103.

Egli, D. B. (2021). Crop management: principles and practices.

Okereke, P. O. (2021). Principles of Crop Production. Agricultural Technology for Colleges, 87.

Unger, P. W., Jones, O. R., & Steiner, J. L. (1988). Principles of crop and soil management procedures for maximizing production per unit rainfall. Drought research priorities for the dryland tropics. ICRISAT, Patancheru, AP, 502324, 97-112.

2

Precision Agriculture for Better Future

Darshana Patra[1], Anyesha Anandita Prusty[2], Swarg Sagar Mohanty[3] and Abhisek Samantaray[4]

[1]Department of Genetics and Plant Breeding, Faculty of Agricultural Sciences, Siksha o' Anusandhan (Deemed to be University), Bhubaneswar Odisha
[2]Department of Agronomy, Siksha o' Anusandhan (Deemed to be University) Bhubaneswar, Odisha
[3]Extension Education and Communication, Siksha o' Anusandhan (Deemed to be University), Bhubaneswar, Odisha
[4]Agronomy, Siksha o' Anusandhan (Deemed to be University) Bhubaneswar, Odisha

Abstract

Precision agriculture represents a transformative approach to farming that harnesses advanced technologies to optimize resource use, enhance productivity, and promote sustainability. This abstract delves into the potential of precision agriculture to shape a better future for agricultural systems worldwide. By leveraging tools such as global positioning systems (GPS), remote sensing, and data analytics, precision agriculture enables farmers to make informed decisions tailored to the specific needs of their crops and fields. This targeted approach enhances efficiency by optimizing inputs such as water, fertilizers, and pesticides, reducing waste, and minimizing environmental impacts. Moreover, precision agriculture facilitates site-specific management practices, allowing farmers to address variability within fields and maximize yields while conserving resources. Beyond productivity gains, precision agriculture promotes environmental stewardship by reducing chemical runoff, soil erosion, and greenhouse gas emissions. Additionally, by improving economic viability and resilience, precision agriculture contributes to the long-term sustainability of agricultural systems, ensuring food security and livelihoods for future generations. Embracing precision agriculture holds the promise of revolutionizing farming practices, empowering farmers, and fostering a more sustainable and resilient agricultural future.

Keywords: *precision, productivity, enhances, minimizing, optimizing*

Introduction

Precision agriculture, also known as precision farming or smart farming, represents a ground breaking approach to agricultural management that integrates cutting-edge technologies and data-driven techniques to optimize crop production. This introduction sets the stage for exploring the transformative potential of precision agriculture in revolutionizing traditional farming practices. By harnessing advancements in sensors, drones, artificial intelligence, and global positioning systems (GPS), precision agriculture empowers farmers to make informed decisions based on real-time data and spatial variability within fields. The overarching goal of precision agriculture is to maximize yields, minimize input costs, and reduce environmental impacts by precisely tailoring farming practices to the specific needs of crops and soil conditions. As global population growth and climate change exert increasing pressure on agricultural systems, the adoption of precision agriculture emerges as a crucial strategy for enhancing food security, promoting sustainability, and ensuring the long-term viability of farming operations. Through this introduction, we embark on a journey to explore the multifaceted dimensions of precision agriculture and its profound implications for the future of agriculture.

Importance in Agriculture

The importance of precision agriculture cannot be overstated in today's rapidly evolving agricultural landscape. Precision agriculture offers a paradigm shift in farming practices, providing farmers with unprecedented tools and insights to optimize resource management, increase productivity, and promote sustainability. One of the key advantages of precision agriculture lies in its ability to enable farmers to make data-driven decisions tailored to the unique variability within their fields. By leveraging technologies such as GPS-guided tractors, drones, and remote sensing, farmers can accurately assess soil properties, monitor crop health, and identify areas of inefficiency or stress. This targeted approach allows for precise application of inputs such as water, fertilizers, and pesticides, minimizing waste and environmental impact while maximizing yield potential. Moreover, precision agriculture holds immense potential for enhancing economic viability by reducing input costs, improving crop quality, and increasing overall farm profitability. As global challenges such as climate change, population growth, and resource scarcity intensify, the adoption of precision agriculture becomes increasingly imperative for ensuring food security, mitigating environmental degradation, and sustaining rural livelihoods. In essence, precision agriculture represents a transformative force that empowers farmers to navigate complex agricultural challenges and unlock the full potential of modern farming practices.

Principles Associated with Precision Agriculture

- Precision agriculture is guided by several key principles that underpin its approach to optimizing farming practices through advanced technologies and data-driven decision-making. Firstly, precision agriculture emphasizes the importance of site-specific management, recognizing that fields exhibit spatial variability in soil properties, nutrient levels, and crop health. By employing tools such as GPS, remote sensing, and soil sensors, farmers can accurately assess this variability and tailor management practices accordingly.
- Secondly, precision agriculture prioritizes data collection and analysis to inform decision-making. Through the use of sensors, drones, and satellite imagery, farmers gather real-time data on soil moisture, crop growth, and pest infestations. Analyzing this data allows for precise identification of areas requiring intervention, whether it be targeted irrigation, fertilizer application, or pest control measures.
- Another principle of precision agriculture is the adoption of variable rate technology (VRT), which enables farmers to apply inputs at variable rates across their fields based on spatial data and crop requirements. This allows for more efficient use of resources, reducing input costs and minimizing environmental impacts.
- Additionally, precision agriculture promotes the integration of automation and robotics to streamline farming operations and improve efficiency. Autonomous vehicles, robotic welders, and drones equipped with spraying systems are examples of technologies that enable precise and timely field operations while reducing labour requirements.
- Finally, precision agriculture embraces the concept of continuous improvement through monitoring, evaluation, and adaptation. By regularly assessing the outcomes of management practices and incorporating feedback from data analysis, farmers can refine their approaches over time to optimize yields, minimize inputs, and enhance sustainability.

Precision Agriculture

Precision agriculture employs a variety of methods and technologies to optimize farming practices and enhance productivity. Here are some key methods used in precision agriculture:

1. **Global Positioning System (GPS) Technology**: GPS technology allows farmers to accurately determine the geographic location of

field boundaries, landmarks, and equipment within their fields. This information is used for precise navigation, mapping, and data collection. Global Positioning System (GPS) technology has revolutionized precision agriculture by providing farmers with accurate and real-time spatial information to enhance farm management practices. With GPS technology, farmers can precisely determine the geographic location of field boundaries, landmarks, and equipment within their fields. This information serves as the foundation for precise navigation, mapping, and data collection, enabling farmers to optimize field operations and resource use. GPS-enabled equipment, such as tractors, planters, and harvesters, can be guided along predetermined paths with centimetre-level accuracy, reducing overlap and minimizing inputs such as fuel, seed, and fertilizer. Furthermore, GPS technology facilitates the creation of detailed field maps, allowing farmers to identify and analyse spatial variability in soil properties, crop health, and yield potential. By integrating GPS technology into precision agriculture systems, farmers can make informed decisions, optimize inputs, and increase productivity while minimizing environmental impacts. In essence, GPS technology plays a pivotal role in driving the adoption and success of precision agriculture, paving the way for more efficient and sustainable farming practices.

2. **Remote Sensing**: Remote sensing involves the use of satellites, drones, or aircraft equipped with sensors to capture images and data about crop health, soil moisture, nutrient levels, and other important parameters. This information is used to monitor field conditions, detect variability, and make informed management decisions. Remote sensing has emerged as a powerful tool in precision agriculture, offering farmers valuable insights into crop health, soil conditions, and environmental factors. By utilizing satellites, drones, or aircraft equipped with sensors, remote sensing technologies capture high-resolution images and data about agricultural fields. These images provide detailed information on crop vigor, stress, and variability, allowing farmers to monitor field conditions and identify potential issues such as nutrient deficiencies, pest infestations, or water stress. Additionally, remote sensing data can be used to create precise maps of field boundaries, soil types, and crop performance, enabling farmers to make informed management decisions tailored to specific areas within their fields. With the ability to collect data quickly and at large scales, remote sensing enhances efficiency in farm management and resource allocation. By integrating remote sensing into precision agriculture systems, farmers can optimize inputs,

minimize losses, and increase yields while promoting sustainability and environmental stewardship. In summary, remote sensing technology is a valuable asset in the arsenal of precision agriculture, empowering farmers with the information they need to make data-driven decisions and optimize crop production.

3. **Variable Rate Technology (VRT):** VRT enables farmers to apply inputs such as fertilizers, pesticides, and irrigation water at variable rates across their fields based on spatial data and crop requirements. This allows for more targeted application, minimizing waste and optimizing resource use. Variable Rate Technology (VRT) is a cornerstone of precision agriculture, offering farmers the ability to apply inputs such as fertilizers, pesticides, and irrigation water at variable rates across their fields based on spatial data and crop requirements. This technology allows farmers to tailor input application to the specific needs of different areas within their fields, maximizing efficiency and optimizing resource use. By utilizing sensors, GPS technology, and data analytics, VRT systems create prescription maps that delineate zones with varying soil types, nutrient levels, and crop conditions. These maps guide the precise application of inputs, ensuring that crops receive the right amount of nutrients, water, and protection at the right time and in the right place. VRT offers several benefits to farmers, including increased yields, reduced input costs, and minimized environmental impacts. By applying inputs only where they are needed and at the appropriate rates, farmers can minimize waste and reduce the risk of over-application, leading to cost savings and improved profitability. Moreover, VRT enables farmers to manage field variability more effectively, addressing soil nutrient deficiencies, pest pressures, and water stress on a site-specific basis. Overall, Variable Rate Technology represents a significant advancement in precision agriculture, empowering farmers to optimize crop production while promoting sustainability and environmental stewardship. By embracing VRT, farmers can achieve greater efficiency, profitability, and resilience in their farming operations, contributing to a more sustainable and productive agricultural future.

4. **Precision Irrigation**: Precision irrigation systems, such as drip irrigation and sprinkler systems equipped with soil moisture sensors, deliver water to crops with high precision, minimizing water waste and optimizing water use efficiency. Precision irrigation is a pivotal component of precision agriculture, offering farmers the ability to deliver water to crops with unprecedented accuracy and efficiency. This technology

utilizes sensors, actuators, and data analytics to precisely monitor and control water application based on real-time data and crop needs. By optimizing water use at a spatial and temporal scale, precision irrigation systems minimize water waste, reduce energy consumption, and improve crop yields. One of the key advantages of precision irrigation is its ability to tailor water application to the specific requirements of different areas within fields. Soil moisture sensors and weather data are used to determine soil moisture levels, crop water uptake rates, and environmental conditions, allowing farmers to apply water precisely where and when it is needed. This targeted approach ensures that crops receive the right amount of water to support growth and development while minimizing water stress and runoff. Precision irrigation systems come in various forms, including drip irrigation, micro-sprinklers, and pivot irrigation systems equipped with variable rate technology. These systems offer flexibility in water application methods and can be adapted to different crop types, soil conditions, and field topographies. Moreover, by integrating precision irrigation with other precision agriculture technologies such as GPS and remote sensing, farmers can optimize irrigation scheduling, minimize water use, and maximize crop productivity.

5. **Automated Machinery**: Automated machinery, including GPS-guided tractors, planters, and harvesters, enable precise field operations such as planting, seeding, spraying, and harvesting. This reduces operator error, improves efficiency, and allows for more timely operations. Automated machinery plays a pivotal role in advancing precision agriculture, offering farmers a range of automated solutions to streamline field operations, enhance efficiency, and improve productivity. From GPS-guided tractors to robotic harvesters, automation technologies are revolutionizing the way farming tasks are performed, reducing labour requirements and increasing operational precision. One of the key benefits of automated machinery is its ability to perform tasks with a high degree of accuracy and consistency. GPS-guided tractors, for example, can follow predefined paths with centimetre-level accuracy, reducing overlap and minimizing input waste. This precision not only improves efficiency but also optimizes resource use, resulting in cost savings and environmental benefits. Moreover, automated machinery enables farmers to operate more efficiently and effectively, especially in large-scale farming operations. Tasks such as planting, seeding, spraying, and harvesting can be performed autonomously or semi-autonomously, freeing up labour resources and allowing farmers to focus on higher-

level decision-making and management activities. Additionally, automation technologies contribute to improved safety and comfort for farm workers by reducing the need for manual labour in strenuous or hazardous tasks. By automating repetitive or dangerous tasks, farmers can minimize the risk of injuries and create a safer working environment for themselves and their employees.

6. **Data Analytics and Decision Support Systems**: Data analytics tools and decision support systems analyse data collected from various sources, such as sensors and remote sensing, to provide farmers with actionable insights and recommendations for optimizing management practices. Data analytics and decision support systems are integral components of precision agriculture, providing farmers with valuable insights and recommendations to optimize farm management practices. These systems harness the power of big data, advanced analytics, and machine learning algorithms to analyse complex datasets and generate actionable information for farmers. One of the key benefits of data analytics in precision agriculture is its ability to process and interpret large volumes of data collected from various sources, including sensors, satellites, and farm equipment. This data may include information on soil properties, crop health, weather conditions, and field operations. By analysing this data, farmers can gain valuable insights into field variability, identify trends and patterns, and make informed decisions to improve crop management practices. Decision support systems build upon data analytics by providing farmers with tools and recommendations to guide their decision-making process. These systems may include crop modelling tools, yield prediction models, and prescription mapping software that help farmers optimize inputs such as fertilizers, pesticides, and irrigation water. By integrating data analytics with decision support systems, farmers can implement precision agriculture practices more effectively, maximizing crop yields while minimizing input costs and environmental impacts. Moreover, data analytics and decision support systems enable farmers to adopt a proactive approach to farm management, rather than a reactive one. By continuously monitoring field conditions and analysing data in real-time, farmers can anticipate potential issues such as pest outbreaks, nutrient deficiencies, or water stress and take pre-emptive action to mitigate risks and optimize crop performance.

7. **Geographic Information Systems (GIS):** GIS software allows farmers to create maps, analyse spatial data, and visualize field variability. GIS

tools are used for precision mapping, zone delineation, and creating prescription maps for variable rate application. Geographic Information Systems (GIS) are indispensable tools in precision agriculture, providing farmers with powerful capabilities to analyse, visualize, and manage spatial data related to their fields. GIS technology enables farmers to create detailed maps that incorporate various layers of information, such as soil types, topography, yield variability, and crop health indicators. By integrating these spatial datasets, farmers gain valuable insights into field variability and can make informed decisions to optimize crop management practices. One of the key benefits of GIS in precision agriculture is its ability to facilitate precision mapping and zone delineation. Farmers can use GIS software to create maps that identify areas of high and low productivity within their fields, allowing them to implement site-specific management practices tailored to the unique characteristics of each zone. This targeted approach enables farmers to optimize inputs such as fertilizers, pesticides, and irrigation water, maximizing yields while minimizing environmental impacts. Moreover, GIS technology enables farmers to analyse spatial relationships and patterns, helping them to identify correlations between field parameters and crop performance. For example, GIS can be used to identify areas of soil compaction, waterlogging, or nutrient deficiency that may be impacting crop growth and yield. By understanding these spatial relationships, farmers can implement targeted interventions to address specific issues and improve overall field productivity. Additionally, GIS facilitates data integration and collaboration across different stakeholders within the agricultural value chain. Farmers can share spatial data with agronomists, consultants, and researchers, enabling collaborative decision-making and knowledge sharing. This collaboration fosters innovation and enables farmers to benefit from collective expertise and insights, ultimately leading to improved farm management practices and productivity. Geographic Information Systems (GIS) play a crucial role in precision agriculture by providing farmers with the tools they need to analyse, visualize, and manage spatial data related to their fields. By harnessing the power of GIS technology, farmers can make informed decisions, optimize resource use, and increase productivity, contributing to a more sustainable and resilient agricultural future.

8. **Robotics and Automation**: Robotics and automation technologies, such as robotic weeders, drones for crop monitoring, and autonomous vehicles for field operations, streamline tasks, reduce labour requirements, and

improve efficiency in agricultural operations. Robotics and automation are revolutionizing precision agriculture by offering farmers innovative solutions to streamline field operations, reduce labour requirements, and improve efficiency. In recent years, advancements in robotics technology have led to the development of autonomous vehicles, robotic systems, and drones that can perform a wide range of tasks in agricultural settings. One of the key benefits of robotics and automation in precision agriculture is their ability to perform repetitive or labour-intensive tasks with precision and consistency. Autonomous vehicles equipped with GPS and sensors can navigate fields and perform tasks such as planting, seeding, spraying, and harvesting with minimal human intervention. This reduces the need for manual labour and allows farmers to allocate resources more efficiently. Moreover, robotics and automation technologies enable farmers to operate more efficiently and effectively, especially in large-scale farming operations. Tasks that were once time-consuming and labour-intensive, such as weeding and crop monitoring, can now be performed autonomously or semi-autonomously by robotic systems. This not only improves efficiency but also reduces operational costs and increases productivity. Additionally, robotics and automation contribute to improved safety and comfort for farm workers by reducing exposure to hazardous tasks and environments. By automating tasks such as spraying pesticides or harvesting crops, farmers can minimize the risk of injuries and create a safer working environment for themselves and their employees. Furthermore, robotics and automation technologies facilitate data collection and analysis, providing farmers with valuable insights into field conditions and crop performance. Drones equipped with sensors and cameras can capture high-resolution images and data about crop health, soil moisture, and pest infestations, allowing farmers to make informed decisions and optimize management practices.

9. **Smart Farming Apps and Platforms:** Smart farming apps and platforms provide farmers with tools for data collection, analysis, and decision-making on mobile devices. These apps may include features for monitoring weather conditions, managing field activities, and accessing agronomic recommendations. Smart farming apps and platforms have emerged as valuable tools in precision agriculture, providing farmers with convenient access to data, analysis tools, and management resources on mobile devices. These apps and platforms leverage technology to streamline farm management practices, optimize resource use, and improve decision-making processes. One of the key benefits of smart farming apps and platforms is their ability to facilitate

real-time monitoring and management of farm operations. Farmers can use these tools to remotely monitor field conditions, track weather forecasts, and receive alerts about potential issues such as pest outbreaks or equipment malfunctions. This enables farmers to stay informed and responsive to changing conditions, ultimately improving productivity and reducing risks. Moreover, smart farming apps and platforms offer a wide range of features and functionalities to support various aspects of farm management. Farmers can use these tools to create field maps, record field observations, track input usage, and manage inventory. Additionally, some apps offer tools for financial management, crop planning, and market analysis, helping farmers make informed decisions and maximize profitability. Furthermore, smart farming apps and platforms enable data integration and collaboration across different stakeholders within the agricultural value chain. Farmers can share data with agronomists, consultants, and researchers, facilitating collaborative decision-making and knowledge sharing. This collaboration fosters innovation and enables farmers to benefit from collective expertise and insights, ultimately leading to improved farm management practices and productivity.

Conclusion

In conclusion, precision agriculture represents a transformative approach to farming that harnesses advanced technologies and data-driven techniques to optimize crop production, enhance sustainability, and promote efficiency. Throughout this discussion, we have explored various components of precision agriculture, including GPS technology, remote sensing, variable rate technology, precision irrigation, automated machinery, data analytics, GIS, robotics, and smart farming apps. Each of these components plays a crucial role in revolutionizing farming practices, enabling farmers to make informed decisions, optimize resource use, and increase productivity while minimizing environmental impacts. The adoption of precision agriculture holds immense promise for addressing the challenges facing modern agriculture, including the need to feed a growing global population, mitigate climate change, and conserve natural resources. By embracing precision agriculture technologies and practices, farmers can enhance their competitiveness, reduce costs, and improve sustainability, paving the way for a more resilient and efficient agricultural future. However, it is important to recognize that the successful implementation of precision agriculture requires more than just access to technology. It also requires investment in infrastructure, education, and training to ensure that farmers have the skills and knowledge needed to effectively utilize these technologies. Moreover, there are challenges related to data privacy, interoperability, and affordability that need to be addressed to enable widespread adoption of precision agriculture. Overall, precision agriculture offers a promising path forward for the agricultural industry, empowering farmers to produce more with less, minimize environmental impacts, and ensure food security for future generations. By continuing to innovate, collaborate, and invest in precision agriculture, we can build a more sustainable and resilient agricultural system that meets the needs of both present and future generations.

References

Bendre, M. R., Thool, R. C., & Thool, V. R. (2015, September). Big data in precision agriculture: Weather forecasting for future farming. In 2015 1st international conference on next generation computing technologies (NGCT) (pp. 744-750). IEEE.

Bhattacharyay, D., Maitra, S., Pine, S., Shankar, T., & Pedda Ghouse Peera, S. K. (2020). Future of precision agriculture in India. Protected cultivation and smart agriculture, 1, 289-299.

Demirbaş, N. (2018). Precision agriculture in terms of food security: Needs for the future. Precision Agriculture, 27(28).

Gyarmati, G. and Mizik, T. (2020). The present and future of the precision agriculture. In 2020 IEEE 15th International Conference of System of Systems Engineering (SoSE) (pp. 593-596). IEEE.

Hedley, C. (2015). The role of precision agriculture for improved nutrient management on farms. Journal of the Science of Food and Agriculture, 95(1), 12-19.

Schepers, J. S., & Francis, D. D. (1998). Precision agriculture—what's in our future. Communications in soil science and plant analysis, 29(11-14), 1463-1469.

Shafi, U., Mumtaz, R., García-Nieto, J., Hassan, S. A., Zaidi, S. A. R., & Iqbal, N. (2019). Precision agriculture techniques and practices: From considerations to applications. Sensors, 19(17), 3796.

3

Soil Management in Agricultural Crops

***P. Kalaiselvi*[1]*, Kushal Sachan*[2]*, Vijay Kumar*[3] *and Rini Labanya*[4]**

[1]*ICAR- Krishi Vigyan Kendra, Tamil Nadu Agricultural University, Salem*
[2]*Department of Soil Science and Agricultural Chemistry, C.S. Azad University of Agriculture and Technology, Kanpur, Uttar Pradesh*
[3]*Plant Breeding, ICAR- Sugarcane Breeding institute, Regional Centre Karnal, Haryana*
[4]*Sri Sri University, Cuttack, Odisha*

Abstract

Soil management is a cornerstone of agricultural crop production, encompassing a range of practices aimed at optimizing soil health, fertility, and productivity. This abstract delves into the significance of soil management in agricultural crops, exploring key principles, challenges, and strategies for sustainable soil stewardship. Effective soil management begins with soil assessment, including soil testing to analyse nutrient levels, pH, and texture, providing crucial insights for fertilizer application and soil amendment. Implementing soil conservation practices, such as minimal tillage, cover cropping, and contour farming, helps prevent soil erosion, maintain soil structure, and preserve soil moisture. Moreover, soil management involves the judicious use of organic matter, such as compost and crop residues, to enhance soil fertility, microbial activity, and carbon sequestration. Challenges such as soil degradation, nutrient depletion, and salinity necessitate proactive soil management approaches, including soil restoration techniques and precision agriculture technologies. By prioritizing soil health and resilience, farmers can optimize crop yields, mitigate environmental impacts, and contribute to the long-term sustainability of agricultural systems. In essence, soil management serves as the foundation for sustainable crop production, highlighting the importance of stewardship and conservation in agricultural practices.

Keywords: *agricultural, texture, soil erosion, tillage*

Introduction

Soil management refers to a set of practices and strategies employed by farmers and land managers to maintain, improve, and sustain the health, fertility, and productivity of soil in agricultural systems. It encompasses a range of activities aimed at optimizing soil conditions for crop growth while minimizing environmental degradation. Soil management practices include soil testing to assess nutrient levels and pH, the application of fertilizers and soil amendments to correct deficiencies, erosion control measures to prevent soil loss, and conservation practices to enhance soil structure and health. Additionally, soil management involves the integration of organic matter, such as compost and crop residues, to enhance soil fertility and microbial activity. The overarching goal of soil management is to ensure the long-term sustainability of agricultural systems by preserving soil health, supporting crop productivity, and minimizing environmental impacts.

Method of Soil Management

1. **Soil Testing and Analysis**: Soil testing and analysis play a pivotal role in soil management for agricultural crops, providing farmers with critical information about soil health, fertility, and nutrient status. By understanding the composition of their soil, farmers can make informed decisions about fertilization, soil amendments, and crop selection to optimize yields and minimize environmental impacts. For example, soil testing can reveal deficiencies or excesses of essential nutrients such as nitrogen, phosphorus, and potassium, enabling farmers to tailor their fertilizer applications accordingly. If a soil test indicates low levels of phosphorus, for instance, farmers can apply phosphorus-rich fertilizers to replenish the nutrient and support healthy crop growth. Conversely, if soil pH is too high or too low, amendments such as lime or sulphur can be applied to adjust pH levels and create optimal growing conditions for crops. Soil testing can also help identify soil characteristics such as texture, organic matter content, and cation exchange capacity, which influence soil fertility and water-holding capacity. By analyzing these factors, farmers can implement soil management practices such as irrigation scheduling, cover cropping, and conservation tillage to improve soil structure and reduce erosion. Furthermore, soil testing can aid in diagnosing soil-related problems such as salinity, compaction, or soil-borne diseases, allowing farmers to implement targeted solutions to mitigate these issues. For example, if soil salinity is identified as a problem, farmers can implement strategies such as improved drainage or salt-tolerant crop varieties to address the issue and maintain crop

productivity. Overall, soil testing and analysis are essential tools for effective soil management in agricultural crops, enabling farmers to make data-driven decisions that promote soil health, optimize crop yields, and sustainably manage natural resources. By regularly testing their soil and acting on the results, farmers can ensure the long-term productivity and viability of their agricultural operations. Soil testing plays a crucial role in optimizing crop production by providing valuable insights into soil nutrient levels, pH, and other properties. Tailoring management practices based on soil test results can help farmers achieve optimal crop yields and quality. Here are some examples of soil testing applications with specific crop examples:

- **Corn (Maize):** Soil testing for corn production typically involves assessing soil nutrient levels, particularly nitrogen (N), phosphorus (P), and potassium (K). Corn has high nutrient demands, especially during the vegetative and reproductive growth stages. Soil tests can help determine the appropriate rates and timing of fertilizer applications to meet these nutrient requirements. For example, a soil test indicating low phosphorus levels may prompt farmers to apply phosphorus-containing fertilizers such as diammonium phosphate (DAP) before planting corn to ensure adequate nutrient availability during early growth stages.

- **Soybeans:** Soybeans have a relatively lower nitrogen requirement compared to other crops due to their ability to fix atmospheric nitrogen through symbiotic relationships with soil bacteria. However, soil testing for soybeans is still essential for assessing phosphorus, potassium, and micronutrient levels. Soil tests may also help identify pH imbalances that could affect nutrient availability. For instance, soybeans are sensitive to low soil pH, so liming materials may be applied based on soil test results to raise pH and improve soybean yields.

- **Wheat:** Soil testing is critical for wheat production to ensure adequate nutrient levels for optimal growth and grain yield. Wheat has high nitrogen requirements, particularly during the early stages of growth. Soil tests can help determine the appropriate nitrogen fertilizer rates and application methods, such as split applications or top-dressing, to meet wheat's nutrient demands while minimizing nitrogen losses through leaching or volatilization. Additionally, soil tests can identify micronutrient deficiencies such as zinc or boron, which can affect wheat quality and yield.

- **Potatoes:** Potatoes are heavy feeders of nutrients, particularly potassium and phosphorus. Soil testing is essential for potatoes to assess nutrient levels and pH, as well as to identify potential soil-borne diseases such as common scab or potato cyst nematodes. Soil tests may prompt farmers to apply fertilizers containing high levels of phosphorus and potassium to meet potato nutrient requirements. Additionally, soil tests can inform pH adjustments, as potatoes prefer slightly acidic soil conditions (pH 5.0-6.0).

2. **Fertilization Practices**: Fertilization practices are a cornerstone of soil management in agricultural crop production, aimed at providing essential nutrients necessary for crop growth, development, and yield optimization. These practices involve the application of fertilizers, both organic and synthetic, to replenish nutrients in the soil and address deficiencies identified through soil testing. The key nutrients typically targeted for replenishment include nitrogen (N), phosphorus (P), potassium (K), and various micronutrients essential for plant health. Nitrogen is crucial for vegetative growth, photosynthesis, and protein synthesis in plants. Fertilization practices often involve applying nitrogen fertilizers to meet crop demands, especially in crops with high nitrogen requirements such as corn, wheat, and rice. Phosphorus plays a vital role in root development, energy transfer, and flowering. Phosphorus fertilizers are applied to soils deficient in phosphorus to ensure optimal crop growth and yield, particularly in crops like potatoes, soybeans, and citrus fruits. Potassium is essential for regulating water uptake, enzyme activation, and overall plant health. Fertilization practices include applying potassium fertilizers to maintain adequate soil potassium levels, benefiting crops such as tomatoes, bananas, and cotton. Additionally, micronutrients such as zinc, iron, manganese, and boron are crucial for various physiological processes in plants. Soil testing helps identify micronutrient deficiencies, allowing farmers to apply targeted fertilizers or foliar sprays to address these deficiencies and ensure optimal crop performance. Fertilization practices also consider factors such as nutrient uptake efficiency, soil characteristics, crop rotation, and environmental sustainability. Proper timing, placement, and rates of fertilizer application are essential to maximize nutrient utilization by crops and minimize environmental impacts such as nutrient runoff, leaching, and greenhouse gas emissions. Moreover, integrating organic fertilizers, cover cropping, and conservation tillage practices can enhance soil fertility, biological activity, and nutrient cycling, reducing reliance on synthetic fertilizers and promoting soil health and sustainability.

3. **Organic Matter Management**: Organic matter management is a foundational aspect of soil management in agricultural crop production, focusing on the integration of organic materials to enhance soil fertility, structure, and overall health. Organic matter, derived from plant and animal residues, serves as a vital source of nutrients and energy for soil microorganisms, fostering nutrient cycling and supporting plant growth. In agricultural crop production, organic matter management involves various practices aimed at increasing soil organic matter content and improving soil quality. These practices include the incorporation of crop residues, cover cropping, composting, and the application of organic amendments such as manure or biochar. One key practice is the incorporation of crop residues into the soil after harvest. Crop residues, such as stalks, leaves, and roots, contain valuable nutrients that can be recycled back into the soil through decomposition. This process not only enriches the soil with essential nutrients but also improves soil structure, water retention, and microbial activity. Cover cropping is another effective organic matter management practice wherein cover crops, such as legumes or grasses, are grown during fallow periods or between cash crops. These cover crops help suppress weeds, prevent soil erosion, and add organic matter to the soil through biomass accumulation and root exudation. As cover crops decompose, they release nutrients, enriching the soil and providing a natural source of fertility for subsequent crops. Composting is a valuable technique for converting organic residues, such as crop residues, kitchen scraps, and yard waste, into nutrient-rich compost. Compost improves soil fertility, enhances microbial activity, and promotes soil aggregation, resulting in healthier soils and improved crop productivity. Additionally, the application of organic amendments such as manure or biochar can further enrich the soil with organic matter and nutrients. These amendments not only provide essential nutrients but also contribute to soil carbon sequestration and climate change mitigation.

4. **Conservation Tillage**: Conservation tillage is a farming practice that aims to reduce soil erosion, improve soil health, and conserve water by minimizing soil disturbance during crop cultivation. Unlike conventional tillage, which involves ploughing and intensive soil disturbance, conservation tillage methods disturb the soil to a lesser extent, preserving soil structure, organic matter, and beneficial soil organisms. In conservation tillage systems, various techniques are employed to minimize soil disturbance, such as no-till, reduced tillage, and strip tillage. No-till farming involves planting crops directly into untilled soil,

leaving the previous crop residue on the soil surface as a protective layer. Reduced tillage practices involve shallow soil cultivation, typically using specialized equipment like chisel ploug or vertical tillage implements. Strip tillage combines elements of both no-till and conventional tillage, with tilled strips where crops are planted interspersed with untilled strips where crop residue remains intact. Conservation tillage offers numerous benefits for agricultural crop production. By reducing soil erosion, conservation tillage helps preserve soil structure and fertility, preventing nutrient loss and maintaining soil health over time. The retention of crop residue on the soil surface provides organic matter, which improves soil structure, increases water infiltration, and reduces surface runoff. Additionally, conservation tillage practices require less fuel, labour, and equipment compared to conventional tillage, resulting in cost savings for farmers and reduced greenhouse gas emissions from agricultural operations. However, successful adoption of conservation tillage requires careful management and adaptation to local soil and climatic conditions. Challenges such as weed control, soil compaction, and pest management may arise in conservation tillage systems, requiring integrated approaches to address these issues effectively. Additionally, the transition to conservation tillage may involve a learning curve for farmers accustomed to conventional tillage practices, necessitating education and support to facilitate adoption and successful implementation.

5. **Crop Rotation and Diversification**: Crop rotation and diversification are important agricultural practices that involve alternating the types of crops grown on a particular piece of land over time. This approach offers numerous benefits for soil health, pest management, nutrient cycling, and overall farm sustainability. Crop rotation involves the sequential planting of different crops in the same field, typically over multiple growing seasons. By rotating crops, farmers can disrupt pest and disease cycles, reduce soil erosion, and improve soil fertility. For example, planting leguminous crops like soybeans or peas in rotation with cereal crops like corn or wheat can help fix atmospheric nitrogen, reducing the need for synthetic fertilizers and enhancing soil fertility. Additionally, rotating crops with different root structures can help break up soil compaction and improve soil structure, promoting better water infiltration and nutrient uptake. Diversification, on the other hand, involves growing a variety of crops within a farming system. Diversified cropping systems can include grains, legumes, vegetables, fruits, and cover crops, among others. This approach offers several benefits,

including risk mitigation against crop failure due to adverse weather conditions, market fluctuations, or pest outbreaks. By diversifying crops, farmers can spread their financial risk and maintain more stable income streams throughout the year. Moreover, diversification contributes to biodiversity conservation, providing habitat and food sources for beneficial insects, birds, and other wildlife. Cover crops, which are typically grown during fallow periods or between cash crops, can serve multiple purposes in a diversified cropping system, including soil erosion control, weed suppression, and soil fertility improvement. For example, planting cover crops like clover or rye can add organic matter to the soil, suppress weeds, and enhance soil health during periods when the main cash crop is not growing.

6. **Irrigation Management**: Irrigation management is an integral component of soil management in agricultural crop production, ensuring that crops receive adequate water to support growth and development while minimizing water waste and environmental impacts. Effective irrigation management involves the strategic application of water based on crop water requirements, soil characteristics, and climatic conditions. Various irrigation methods are utilized in agricultural crop production, including surface irrigation, sprinkler irrigation, drip irrigation, and subsurface irrigation. Each method has its advantages and limitations, and the choice of irrigation system depends on factors such as crop type, soil texture, topography, and water availability. Surface irrigation, such as furrow or flood irrigation, involves the application of water directly to the soil surface, allowing it to infiltrate and move through the root zone. While surface irrigation is relatively simple and cost-effective, it can result in water loss through runoff and evaporation, as well as uneven water distribution. Sprinkler irrigation systems distribute water through overhead sprinklers, covering the crop area with a uniform layer of water. Sprinkler irrigation is suitable for a wide range of crops and soil types and allows for precise control of water application. However, sprinkler systems can be prone to water losses due to wind drift and evaporation, particularly in windy or hot conditions. Drip irrigation, also known as trickle irrigation, delivers water directly to the root zone of plants through a network of drip lines or emitters. Drip irrigation is highly efficient, minimizing water waste and soil erosion while promoting better nutrient uptake and crop growth. This method is particularly well-suited for row crops, orchards, and greenhouse production systems. Subsurface irrigation systems deliver water below the soil surface, either through buried drip lines or perforated tubes. Subsurface irrigation minimizes

water loss due to evaporation and runoff and reduces weed growth and soil compaction. However, subsurface irrigation systems require careful management to prevent clogging of emitters and ensure uniform water distribution throughout the root zone.

7. **Cover Cropping**: Cover cropping is a sustainable soil management practice widely used in agricultural crop production to improve soil health, fertility, and resilience. It involves planting a cover crop, typically a non-cash crop, between periods of cash crop cultivation to cover and protect the soil. Cover crops provide numerous benefits to the soil and the overall cropping system. One of the primary benefits of cover cropping is soil erosion control. The dense root systems and aboveground biomass of cover crops help to anchor the soil, reducing erosion caused by wind and water. This protects the soil from nutrient and topsoil loss, preserving its structure and fertility. Additionally, cover crops contribute to soil fertility by fixing nitrogen from the atmosphere, increasing organic matter content, and enhancing nutrient cycling. Leguminous cover crops, such as clover, peas, or vetch, have the ability to form symbiotic relationships with nitrogen-fixing bacteria, thereby enriching the soil with nitrogen—a vital nutrient for plant growth. As cover crops decompose, they release nutrients into the soil, making them available for subsequent cash crops. Cover cropping also improves soil structure and tilth. The root systems of cover crops help to break up compacted soils, improve water infiltration and retention, and create channels for air and water movement. This enhances soil aeration and drainage, promoting a healthy soil environment for crop roots and soil organisms. Furthermore, cover crops contribute to weed suppression by shading the soil surface and competing with weeds for nutrients, water, and light. This reduces the need for synthetic herbicides and tillage, minimizing soil disturbance and preserving soil structure.

8. **Mulching Practices**: Mulching is a widely practiced soil management technique in agricultural crop production that involves covering the soil surface with a layer of organic or synthetic material. This layer of mulch provides numerous benefits to the soil, crops, and overall cropping system. One of the primary benefits of mulching is soil moisture conservation. By covering the soil surface, mulch helps to reduce evaporation, retain soil moisture, and maintain consistent soil moisture levels, particularly during periods of drought or high temperatures. This helps to mitigate water stress in crops, promote healthy root development, and improve overall water use efficiency. Additionally, mulching helps to control soil

temperature by insulating the soil surface from extreme temperature fluctuations. In hot climates, mulch helps to keep the soil cool, reducing heat stress on plant roots and preserving soil microbial activity. In colder climates, mulch acts as an insulating layer, helping to moderate soil temperatures and protect crops from frost damage. Mulching also suppresses weed growth by blocking sunlight and preventing weed seeds from germinating. This reduces the need for manual weeding or chemical herbicides, minimizing soil disturbance and preserving soil structure. Weed suppression through mulching helps to maintain a clean and weed-free growing environment for crops, reducing competition for nutrients, water, and light. Furthermore, mulching contributes to soil fertility and health by adding organic matter to the soil as the mulch material decomposes. Organic mulches, such as straw, leaves, or compost, gradually break down over time, releasing nutrients into the soil and improving soil structure. This enhances soil fertility, promotes microbial activity, and increases nutrient cycling, benefiting crop growth and productivity.

9. **Soil Amendments**: Soil amendments are materials added to soil to improve its physical, chemical, and biological properties, thereby enhancing soil fertility, structure, and overall health. These amendments can include both organic and inorganic materials, such as compost, manure, lime, gypsum, and biochar, among others. Soil amendments play a crucial role in soil management in agricultural crop production by addressing soil deficiencies, enhancing nutrient availability, and promoting optimal crop growth. One of the primary benefits of soil amendments is the improvement of soil fertility. Organic amendments, such as compost and manure, are rich in organic matter and nutrients, including nitrogen, phosphorus, and potassium. When incorporated into the soil, these amendments gradually release nutrients, providing a sustainable source of fertility for crops and promoting microbial activity. Inorganic amendments, such as lime or gypsum, can also improve soil fertility by adjusting soil pH levels and enhancing nutrient availability. Soil amendments also contribute to soil structure improvement by increasing soil organic matter content and promoting aggregation. Organic amendments help to bind soil particles together, improving soil porosity, water infiltration, and root penetration. This enhances soil aeration, drainage, and root development, creating a favourable growing environment for crops and reducing the risk of soil erosion. Furthermore, soil amendments can play a role in soil pH management. Lime is commonly used to raise soil pH in acidic soils, while elemental

sulphur can lower soil pH in alkaline soils. By adjusting soil pH levels, soil amendments help to optimize nutrient availability and uptake by crops, ensuring balanced nutrient management and healthy crop growth. Additionally, certain soil amendments, such as biochar, have the potential to sequester carbon and mitigate greenhouse gas emissions, contributing to climate change mitigation efforts. Biochar is a carbon-rich material produced from biomass through pyrolysis, and when applied to soil, it can enhance soil fertility, water retention, and carbon storage.

10. **Buffer Strips:** Buffer strips are vegetated areas established along the edges of fields or water bodies to protect soil and water resources from the potential impacts of agricultural activities. These strips serve as natural barriers that help reduce soil erosion, filter out sediment and pollutants, and promote biodiversity in agricultural landscapes. One of the primary functions of buffer strips is erosion control. By intercepting runoff from fields, buffer strips help slow down water flow, allowing sediment and suspended particles to settle out before reaching water bodies. This helps prevent soil erosion, sedimentation of waterways, and the loss of valuable topsoil from agricultural fields. Additionally, buffer strips help stabilize stream banks and protect against the scouring effects of floodwaters, reducing the risk of streambank erosion and channel degradation. Buffer strips also play a crucial role in water quality management by filtering out pollutants and nutrients from agricultural runoff. The dense vegetation in buffer strips acts as a natural filter, trapping sediment, nutrients, pesticides, and other contaminants before they can reach nearby water bodies. This helps improve water quality, reduce the risk of nutrient loading and algal blooms, and protect aquatic ecosystems from the harmful effects of pollution. Furthermore, buffer strips provide habitat and refuge for a variety of wildlife species, including birds, insects, and small mammals. The vegetative cover and diverse plant species in buffer strips offer food, shelter, and breeding sites for wildlife, promoting biodiversity and ecosystem resilience in agricultural landscapes. This contributes to the conservation of native species, enhances ecosystem services, and fosters a healthier and more balanced agroecosystem.

11. **Precision Agriculture Technologies**: Precision agriculture technologies revolutionize soil management in agricultural crop production by providing farmers with advanced tools and data-driven insights to optimize soil health, fertility, and productivity. These technologies utilize a combination of sensors, GPS, data analytics, and automated

machinery to precisely monitor and manage soil conditions, nutrient levels, irrigation, and crop growth in real-time. key aspect of precision agriculture technologies is soil mapping and monitoring. High-resolution soil mapping technologies, such as electromagnetic induction (EMI) and soil conductivity sensors, allow farmers to accurately assess soil properties, including texture, moisture content, pH, and nutrient levels, across their fields. This information enables farmers to create detailed soil maps, identify variability within fields, and make informed decisions regarding soil management practices such as nutrient application, irrigation scheduling, and tillage operations. Another critical component of precision agriculture technologies is variable rate application (VRA) systems. VRA systems utilize data from soil maps, crop sensors, and remote sensing technologies to dynamically adjust the application rates of fertilizers, pesticides, and other inputs based on spatial variability within fields. By applying inputs at variable rates tailored to specific soil and crop conditions, farmers can optimize resource use efficiency, minimize input costs, and reduce environmental impacts such as nutrient runoff and leaching. Furthermore, precision agriculture technologies enable farmers to implement site-specific soil management practices, such as precision planting and variable depth tillage. Precision planting systems utilize GPS and automated machinery to precisely place seeds at optimal spacing and depth, maximizing crop uniformity and emergence. Variable depth tillage systems adjust tillage depth and intensity based on soil conditions, reducing soil compaction, erosion, and fuel consumption while preserving soil structure and health. Additionally, precision agriculture technologies facilitate the integration of remote sensing data, such as satellite imagery and aerial drones, to monitor crop health and soil moisture levels. These remote sensing technologies provide farmers with valuable insights into crop growth, stress, and yield potential, allowing for timely interventions and informed decision-making regarding irrigation scheduling, pest management, and harvest planning.

Conclusion

In conclusion, soil management is a fundamental aspect of agricultural crop production, with significant implications for soil health, crop productivity, and environmental sustainability. Through various soil management practices such as crop rotation, cover cropping, mulching, soil amendments, buffer strips, and precision agriculture technologies, farmers can improve soil fertility, structure, and resilience while minimizing soil erosion, nutrient loss, and environmental impacts. By adopting sustainable soil management practices, farmers can enhance soil health, conserve natural resources, and promote the long-term sustainability of agricultural systems. Furthermore, integrating soil management practices into agricultural production systems can help address global challenges such as food security, climate change, and biodiversity conservation. It is imperative for farmers, researchers, policymakers, and stakeholders to collaborate and prioritize sustainable soil management practices to ensure the continued productivity and resilience of agricultural systems for future generations.

References

Karlen, D. L., Eash, N. S., & Unger, P. W. (1992). Soil and crop management effects on soil quality indicators. American Journal of Alternative Agriculture, 7(1-2), 48-55.

Kassam, A. H., Basch, G., Friedrich, T., Shaxson, F., Goddard, T., Amado, T. J., ... & Mkomwa, S. (2014). Sustainable soil management is more than what and how crops are grown. In Rolul agriculturii în acordarea serviciilor ecosistemice şi sociale (pp. 230-270).

Lal, R. (2000). Soil management in the developing countries. Soil Science, 165(1), 57-72.

Ludwig, B., Geisseler, D., Michel, K., Joergensen, R. G., Schulz, E., Merbach, I., ... & Liu, X. (2011). Effects of fertilization and soil management on crop yields and carbon stabilization in soils. A review. Agronomy for Sustainable Development, 31, 361-372.

Mitchell, J., Gaskell, M., Smith, R., Fouche, C., & Koike, S. T. (2000). Soil management and soil quality for organic crops.

Powlson, D. S., Gregory, P. J., Whalley, W. R., Quinton, J. N., Hopkins, D. W., Whitmore, A. P., ... & Goulding, K. W. (2011). Soil management in relation to sustainable agriculture and ecosystem services. Food policy, 36, S72-S87.

Shah, F., & Wu, W. (2019). Soil and crop management strategies to ensure higher crop productivity within sustainable environments. Sustainability, 11(5), 1485.

4

Soil Management in Horticultural Crops

Alan Evon Singh, Sreyansh Singh Chauhan, Ajinkya Ajay Telgote and Pankaj Meena

Department of Horticulture (Fruit Science), Naini Agricultural Institute SHUATS, Prayagraj, Uttar Pradesh, India

Abstract

Soil management in horticultural crops is a critical aspect of sustainable agriculture, essential for maintaining soil fertility, structure, and health to support optimal plant growth and yield. This abstract aims to explore key soil management practices specific to horticultural crops, including fruit trees, vegetables, ornamentals, and herbs. Effective soil management strategies such as cover cropping, mulching, composting, and precision irrigation are discussed, emphasizing their role in enhancing soil fertility, water retention, and nutrient availability while reducing erosion and weed pressure. Additionally, the importance of soil testing and analysis in guiding soil management decisions and addressing nutrient deficiencies is highlighted. The abstract also underscores the significance of integrating organic practices and conservation techniques to promote soil health and sustainability in horticultural crop production. Overall, this abstract provides insights into the importance of soil management practices tailored to the unique needs of horticultural crops, emphasizing their role in ensuring productive and resilient agricultural systems.

Keywords: *soil management, effective, strategies, cover, organic*

Introduction

Soil management in horticultural crops refers to the deliberate and systematic approach of caring for soil to optimize its health, fertility, structure, and overall quality to support the growth and development of horticultural plants. It involves a range of practices aimed at preserving and enhancing soil resources, including soil testing, nutrient management, irrigation scheduling, erosion control, and organic matter addition. The goal of soil management in horticultural crops is to create and maintain an environment in which plants can thrive, maximizing crop yields, quality, and sustainability while minimizing environmental impacts such as soil erosion, nutrient runoff, and soil degradation. By implementing effective soil management strategies, horticultural growers can ensure the long-term productivity and viability of their farming operations, while also promoting soil health and resilience in agricultural ecosystems.

Current Soil Conditions

A study conducted in 2016 over 75 fields revealed that soil compaction was present in 70% of annual crops and 60% of perennial crops. Most of the soils used for both annual and perennial cropping were in a moderate condition. However, the soil condition for established crops was improved by performing cultivations before planting. Nevertheless, some soil formations exhibited instability, especially on soils with a lighter texture and a low content of organic matter. The majority of soil cultivations successfully alleviated compaction, however some cultivations had little impact or even exacerbated compaction due to spreading, puddling, or compression when performed in damp or wet field conditions. Tillage pans may greatly diminish productivity and overall efficiency of production by impeding drainage and root development. Approximately 60% of fields cultivating annual crops and 50% of areas cultivating perennial crops had a well-established tillage pan. Approximately 10% of fields underwent deep cultivation without any evidence of soil compaction, indicating a lack of consideration for soil conditions and the need for subsoiling. The abundance of earthworms in the topsoil may serve as a valuable measure of soil health, and is primarily influenced by the intensity of farming practices and the presence, nutritional value, and consistency of earthworm food sources. Earthworm populations exhibited a general scarcity in both annual and perennial crops, while they were relatively more plentiful in areas with ample crop waste and apple orchards. The regular tilling of the soil is believed to be a significant contributing factor to the very low population of earthworms found in places where crops are grown every year. Fields that had a mustard cover crop reported the greatest populations of earthworms.

Importance of Soil Structure

To preserve soil structure, it is important to prevent soil compaction. This may be accomplished by implementing strategies such as using low ground pressure (LGP) tires, managing wheel loads, and reducing traffic. LGP tires have the ability to enhance traction and fuel efficiency, minimize topsoil compaction, and decrease the breadth of the affected region. Nevertheless, in moist circumstances, even LGP tires may induce wheel slippage and compaction, which hampers the development of crops. Implementing tracks on harvest machines may mitigate compaction depth, however shallow compaction may still have a notable impact. Both the weight of equipment and its utilization are crucial elements to take into account. Controlled traffic farming (CTF) aims to limit compaction to a small area of permanent traffic lanes by implementing stricter guidelines for the use of route ways and tramlines. For effective CTF farming, it is necessary to make adjustments to the equipment in order to ensure that all machines are operating on a uniform track gauge. Seasonal CTF (SCTF) is a method that involves using commonly used tracks and operating widths for most equipment throughout the harvest season. In SCTF systems, tillage is necessary to regulate the compaction effects caused by harvest traffic in the crop growing zone. Manual CTF may be implemented without the use of guidance devices. In this approach, the driver operates the tractor along the same tracks every year, with 'A-B' lines drawn in the hedge or fence line. RTK guidance may attain a high level of precision, with pass-to-pass and static accuracy of ±1–2cm. This allows for precise following of CTF lanes across several years. By using correction from a geostationary satellite orbiting above the equator, it is possible to attain a pass-to-pass precision of ±2cm and an absolute or repeatable positioning accuracy of ±4cm.

Improving Soils

Soil compaction may be enhanced by gradually increasing the amount of organic matter via the use of bulky organic materials, cover crops, green manures, and grass leys over a period of many years. If there are evident indications of compaction, it will be essential to carry out suitable cultivations before plants may effectively carry out their functions. Efficient production on clay and slowly permeable soils relies heavily on field drainage, regardless of the quality of the topsoil structure. To determine if there is a compact layer in the higher subsoil that may be enhanced by subsoiling, it is advisable to excavate a hole and evaluate the soil. Subsoiling should only be performed when the soil at the desired depth is dry and crumbly, allowing it to break apart rather than get compacted. Inspect the soils at the beginning of the process to verify that effective shattering is taking place. In order to prevent soil structural damage, it is important for both the soil surface and the compacted layer to be dry. Winged subsoilers are more efficient than conventional subsoilers in breaking up the soil, however they need more energy to pull. They disrupt a larger amount of soil, often two to three times more, compared to conventional subsoilers. Utilizing leading tines may lead to a greater amount of dirt being disrupted without requiring more force. For optimal results, it is recommended to use a depth wheel or back packer roller to provide a consistent tine depth. The desired tine depth should range from 25-50mm below the base of the compacted layer, with a maximum depth of about 450mm below ground level. Avoid excessive cultivation depth, as it may increase the draft force need by four times and treble the fuel consumption. The spacing between tines is crucial. For conventional subsoilers, the spacing should be up to 1.5 times the depth of the tine. For winged subsoilers, the spacing should be up to 2 times the depth of the tine. And for leading shallow tines, the spacing should be up to 2.5 times the depth of the tine. Following a preliminary test, make necessary modifications to the gaps between objects in order to get the intended level of disruption to the soil.

Organic Content for Horticultural Crops

In order to preserve or improve soil organic matter, it is advisable to frequently apply voluminous organic manures such as green compost or farmyard manure. Green compost, with a total nitrogen content of 250kg/ha, contributes about 4.5 tonnes/ha of organic matter. Within Nitrate Vulnerable Zones (NVZs), it is permissible to apply PAS100 certified green or green/food compost at a doubled rate of up to 500kg total N/ha every 2 years. Orchards cultivating fruit from the Malus, Prunus, or Pyrus genera may receive a maximum of 1,000kg of nitrogen per hectare every 4 years. The addition of organic materials may enhance drainage, soil durability, irrigation effectiveness, crop productivity, and quality. Composts possess a greater amount of lignin, which renders them more resilient to microbial decomposition. Farmyard manures possess a higher amount of recently produced organic material and are more effective in promoting biological processes and enhancing the population of microorganisms.

Cover Crops

Cultivating crops during periods of land inactivity may enhance soil composition within a year and promote the accumulation of organic matter over extended periods of time (10-20 years). Grass leys maintain vegetation coverage year-round, accumulating carbon in both leaves and roots during periods of vigorous development. In addition, they cultivate a compact root system and a layer of 'thatch', which provides a habitat for beneficial organisms that prey on agricultural pests. Cover crops provide winter protection to decrease weed competition, mitigate erosion, surface run-off, and nitrate leaching. They may be used as a component of a comprehensive plan to enhance soil quality and provide other agronomic advantages, such as disrupting pest and disease cycles or integrating into broader weed control programs. Cover crops may also provide broader ecological benefits, creating possibilities for livestock and providing additional food sources for animals. Vegetable rotations often include cover crops, which serve as catch crops during the winter and help improve soil fertility. Fast-growing crops such as cereal rye, ryegrass, and vetches effectively compete with weeds. It is important to develop cover crops that have survived the winter by mid-September in order to effectively cover the land and minimize the loss of nitrates via leaching. Crops that are destroyed at a later stage, with a greater stem size and a more extensive and compact root system, are more likely to contribute to the accumulation of soil organic matter. An extensively cultivated green manure has the potential to yield 1 to 3 metric tons per hectare of organic matter, while a standard application of organic manure typically yields 4 to 5 metric tons per hectare. Organic manures, especially compost, are expected to retain a greater amount of carbon in the soil compared to green manures. Cover crops may be used inside Ecological Focus Areas as part of the Basic Payment Scheme 2018 or as an option under the Countryside Stewardship program.

Measuring and Controlling the Variability of the Soil

Keeping the amount and quality of their crops consistent is a primary concern for farmers. However, the variety in crop development that occurs within the field is often seen in the majority of fields. Variability in the characteristics of the soil is one of the primary variables that determines variances in crop development.

Crop performance will be impacted by a variety of factors, including variations in soil texture that contribute to changes in moisture-holding capacity, organic matter content, nutrient content, drainage, compaction, and soil depth. In situations where there is a significant amount of soil variability that can be handled at practical scales, it may be beneficial to use multiple management regimes within the field, such as for the application of fertilizer or lime, or for cultivation projects. The borders between different kinds of soil may be determined via the use of soil mapping, and field areas can be characterized according to the pH or nutrient indices of the soil. An increasing number of farmers are mapping the variability of the soil as the first step toward comprehending and effectively managing the variability of both the soil and the crops.

Soil Electrical Conductivity (EC)

It is possible to determine changes in soil texture, moisture content, and organic matter content by determining the soil's electrical conductivity (EC), which is a measurement of the soil's capacity to conduct an electric current. Soil electrical conductivity (EC) scanners, which make contact with the soil and detect variations in electrical conductivity throughout the field, and non-contact electromagnetic induction (EMI) sensors, which are held above the soil while scanning, are the two primary kinds of sensors. Texture, moisture, the amount of organic matter present, and bulk density are the primary factors that contribute to the variability of soil electrical conductivity (EC). In contrast, sandy soils with limited particle-to-particle contact and little moisture-holding capacity are poor conductors. Clay soils, on the other hand, have a high moisture-holding capacity and a high particle-to-particle contact, making them very conductive. The observed EC values may be influenced by the moisture content of the soil, but the pattern of fluctuation will not be affected by this factor. Regardless of the amount of moisture present in the soil at the moment of measurement, a soil EC map will always be able to identify places that have a varied soil texture. EC and EMI surveys are carried out at times when there is no crop cover, which is often during the period of time between crops that occurs in the fall and winter. When it comes to 'contact' EC sensors, it is essential to guarantee that there is enough contact between the soil and the coulter. As a result, scans are often performed after harvest, prior to cultivations. A tractor or vehicle is used to drag the instruments across the field at bout widths that are normally between 12 and 24 meters. The information obtained from the sensors is then integrated with GPS data to build a soil electrical conductivity map. There are EC and EMI scanners that are capable of measuring conductivity for two depths simultaneously, which allows them to provide EC maps for both a shallow and a deep vertical cross section. Deep EC data should be used for comparing soil EC and crop yield maps or as a foundation for variable-rate seeding. Shallow EC measurements should be used for guiding soil sample, whereas deep EC measurements should be used for such comparisons.

Satellites Imagining

The amount of sunlight that is reflected by a bare soil surface is referred to as the soil's brightness. This brightness is affected by other elements such as the soil's moisture content, organic matter content, and texture. These surveys are often more cost-effective than soil EC/EMI studies since they are generated from optical satellite images. The analysis of a single satellite picture combined with the application of uniform brightness bandings over the holding is what is used to produce soil brightness maps. It is not possible to compare them to other farms or photographs taken at various times since they are relative and cannot be used for that purpose. In order to evaluate the brightness of the soil, measurements are often performed prior to the establishment of the crop. Each picture displays a slightly varied colour spectrum due to the cultivation technique, the amount of time spent collecting the data, the amount of soil moisture, and the interference from the stump. When it comes to justifying alternative management regimes, soil brightness maps may be of great use in identifying borders between different kinds of soil or circumstances.

Challenges in Soil Management

While soil management in horticultural crops offers numerous benefits, it also presents several challenges that growers may face. Some of the key challenges include:

1. **Nutrient Imbalance:** Achieving and maintaining optimal nutrient levels in the soil can be challenging, particularly in intensive horticultural cropping systems where nutrient uptake rates may vary among different crops. Nutrient imbalances can lead to deficiencies or excesses, affecting plant growth, yield, and overall crop quality.

2. **Soil Erosion:** Soil erosion is a significant challenge in horticultural crop production, particularly on sloping terrain or in areas with intense rainfall. Erosion can result in the loss of valuable topsoil, nutrients, and organic matter, leading to reduced soil fertility and productivity over time.

3. **Soil Compaction:** Heavy machinery and equipment used in horticultural operations can cause soil compaction, particularly in fine-textured soils. Soil compaction restricts root growth, reduces water infiltration and drainage, and impairs soil aeration, leading to decreased crop yields and quality.

4. **Soil Salinity and Alkalinity:** Soil salinity and alkalinity can pose challenges in horticultural crop production, especially in arid and semi-arid regions where irrigation water may contain high levels of salts. High soil salinity and alkalinity can negatively impact plant growth, reduce water uptake, and limit crop yields.

5. **Soil pH Management:** Maintaining optimal soil pH levels is essential for horticultural crop production, as soil pH affects nutrient availability, microbial activity, and overall plant health. Soil pH can be influenced by factors such as soil type, irrigation water quality, and cropping practices, making pH management a complex challenge for growers.

6. **Weed Pressure:** Weeds compete with horticultural crops for nutrients, water, and sunlight, reducing crop yields and quality. Weed management in horticultural crops can be challenging due to the diverse range of weed species, herbicide resistance, and labour-intensive manual weeding requirements.

7. **Soil Health Decline:** Intensive horticultural cropping systems can lead to soil health decline over time due to soil erosion, nutrient depletion, and loss of organic matter. Soil health decline can result in decreased soil fertility, reduced microbial activity, and increased susceptibility to pests and diseases.

8. **Climate Change Impacts:** Climate change poses additional challenges to soil management in horticultural crops, including changes in temperature, precipitation patterns, and extreme weather events. These changes can impact soil moisture levels, nutrient cycling, and overall soil health, affecting crop growth and productivity.

Benefits of Soil Management

Soil management in horticultural crops offers a range of benefits that contribute to improved plant growth, yield, and overall agricultural sustainability. Some of the key benefits include:

1. **Enhanced Soil Fertility:** Effective soil management practices such as soil testing, nutrient management, and organic matter addition help maintain optimal nutrient levels in the soil, promoting healthy plant growth and higher crop yields.

2. **Improved Soil Structure:** Soil management techniques such as cover cropping, reduced tillage, and organic mulching help improve soil structure by enhancing aggregation, reducing compaction, and increasing soil porosity. This improves water infiltration, root penetration, and nutrient uptake by plants.

3. **Increased Water Efficiency:** Precision irrigation methods and soil moisture management practices help optimize water use efficiency in horticultural crops, reducing water waste, minimizing waterlogging or drought stress, and ensuring adequate moisture levels for plant growth.

4. **Weed Suppression:** Mulching and cover cropping practices help suppress weed growth by blocking sunlight, reducing weed seed germination, and competing for resources such as nutrients and water. This reduces the need for herbicides and manual weeding, saving time and labour costs.

5. **Erosion Control:** Soil management practices such as mulching, cover cropping, and contour planting help reduce soil erosion by protecting the soil surface from raindrop impact, runoff, and wind erosion. This helps preserve topsoil, retain soil nutrients, and maintain soil fertility.

6. **Pest and Disease Management:** Integrated pest management (IPM) practices, including crop rotation, biological control, and habitat manipulation, help reduce pest and disease pressure in horticultural crops. Healthy soils with balanced nutrient levels and diverse microbial communities also contribute to plant resilience against pests and diseases.
7. **Environmental Sustainability:** Sustainable soil management practices promote environmental stewardship by minimizing the use of synthetic inputs, reducing soil erosion and nutrient runoff, and conserving water resources. Healthy soils also sequester carbon and contribute to climate change mitigation efforts.
8. **Improved Crop Quality:** Healthy soils with balanced nutrient levels and optimal soil structure provide an ideal growing environment for horticultural crops, resulting in improved crop quality, flavour, colour, and nutritional value.

Conclusion

In conclusion, soil management in horticultural crops is essential for sustaining agricultural productivity, preserving soil health, and promoting environmental sustainability. While horticultural crops offer numerous benefits, including high yields and nutritional value, they also present unique challenges in soil management, such as nutrient imbalances, soil erosion, compaction, and weed pressure. However, by implementing effective soil management practices such as soil testing, nutrient management, erosion control, and conservation tillage, growers can overcome these challenges and optimize soil health and productivity. Sustainable soil management not only benefits crop yields and quality but also contributes to environmental conservation by reducing soil erosion, nutrient runoff, and greenhouse gas emissions. Moving forward, it is imperative for growers, researchers, policymakers, and stakeholders to collaborate and prioritize sustainable soil management practices in horticultural crop production. By investing in soil health and resilience, we can ensure the long-term viability of agricultural systems and support food security, environmental sustainability, and economic prosperity for future generations.

References

Alburquerque, J. A., De la Fuente, C., Campoy, M., Carrasco, L., Nájera, I., Baixauli, C., ... & Bernal, M. P. (2012). Agricultural use of digestate for horticultural crop production and improvement of soil properties. European Journal of Agronomy, 43, 119-128.

del Moral Torres, F. (2024). Special Issue "Horticultural Plant Nutrition, Fertilization and Soil Management". Horticulturae, 10(5), 456.

Fereres, E., Goldhamer, D. A., & Parsons, L. R. (2003). Irrigation water management of horticultural crops. HortScience, 38(5), 1036-1042.

Ganeshamurthy, A. N., Kalaivanan, D., Selvakumar, G., & Panneerselvam, P. (2015). Nutrient management in horticultural crops. Indian Journal of Fertilisers, 11(12), 30-42.

Gautam, R., Pratibha, P. B., & Kumar, A. Advances in Soil Management for Horticulture. Technologies of Horticulture Sciences, 29.

Pradeepkumar, T. (2008). Management of horticultural crops (Vol. 11). New India Publishing.

Vista, S., & Gaihre, Y. (2021, February). Fertilizer Management for Horticultural Crops Using Digital Soil Maps. In Proceedings of the Tenth Natinal Horticulture Workshop, Kathmandu, Nepal (Vol. 28).

5

Integrated Nutrient Management in Agricultural Crops

P. Kalaiselvi[1], Om Prakash[2], Rini Labanya[3] and Sapna[4]

[1]*ICAR- Krishi Vigyan Kendra, Tamil Nadu Agricultural University, Salem*
[2]*Agronomy, Integral Institute of Agricultural Science and Technology (IIAST), Integral University, Lucknow*
[3]*Sri Sri University, Cuttack, Odisha*
[4]*Agricultural Extension Education, SVPUAT, Meerut, Uttar Pradesh*

Abstract

Integrated Nutrient Management (INM) is a comprehensive approach to managing soil fertility and plant nutrition in agricultural crops. This abstract explores the principles, benefits, and challenges associated with INM. INM involves the judicious combination of organic and inorganic nutrient sources, including chemical fertilizers, organic amendments, green manures, and bio fertilizers, to optimize nutrient availability, improve soil health, and enhance crop productivity. By integrating multiple nutrient sources and management practices, INM aims to achieve sustainable and balanced nutrient management, minimize nutrient losses, and reduce environmental impacts such as nutrient runoff and soil degradation. This abstract highlights the importance of INM in addressing nutrient deficiencies, optimizing nutrient use efficiency, and promoting soil fertility and crop sustainability. Despite its potential benefits, the implementation of INM faces challenges such as resource constraints, lack of awareness, and technical know-how. However, with proper planning, training, and support, INM has the potential to revolutionize nutrient management practices in agricultural crops, leading to improved yields, profitability, and environmental stewardship.

Keywords: *health, soil, nutrient, environmental, organic*

Introduction

The term "Integrated Nutrient Management" refers to the process of ensuring that the fertility of the soil and the delivery of plant nutrients are kept at an optimal level in order to maintain the desired level of production. This is accomplished by using an integrated approach to maximize the advantages that may be obtained from all conceivable sources of organic, inorganic, and biological components. Due to the fact that it supplies plants with anchoring, water, and nutrients, soil is an essential component for crop development. It is common for soils to have a certain amount of organic and mineral nutrient sources; but, in order to improve plant development, it is often necessary to supplement these soils with external treatments, sometimes known as fertilisers. It is common practice to apply fertilizers in order to boost the fertility of the soil, encourage the development of plants, increase crop yields, and provide support for agricultural intensification. Generally speaking, fertilizers may be categorized as either organic or mineral. Manure, compost, seaweed, and cereal straw are all examples of organic fertilizers. Organic fertilizers are created from components that are either plant or animal based. In general, organic fertilizers have lower amounts of plant nutrients because they are coupled with organic matter, which enhances the soil's physical and biological features. This results in the soil having better qualities. Mineral fertilizers that are based on nitrogen, potassium, and phosphate are the ones that are used the most often. According to the International Food Policy Research Institute (1995), in order to satisfy the ever-increasing demand for food throughout the world, it will be of the utmost significance to make the most efficient and well-balanced use of nutrient inputs from mineral fertilizers. Since 1960, the usage of mineral fertilizer has expanded by approximately fivefold, which has greatly supported the expansion of the world population. According to Smil (2002), nitrogen-based fertilizer has contributed an estimated forty percent to the gains in per-capita food production over the previous fifty years. In spite of this, mineral fertilizers should not be the only means by which crop nutrient needs are satisfied. This is because environmental issues and economic limits make this situation impossible. As a result, Integrated Nutrient Management has to be used in order to encourage the efficient utilization of all nutrient sources, which includes organic sources, recyclable wastes, mineral fertilizers, and biofertilizers (Roy et al, 2006).

The goal of Integrated Nutrient Management (INM) is to boost crop output while preserving soil productivity for future generations (FAO, 1995a). This is accomplished by integrating the use of several types of soil nutrients, including those that are natural and those that are man-made. On the basis of crop rotation or cropping systems, integrated nutrition management (INM) seeks to maximize the use of nutritional sources rather than concentrating nutrition management strategies on a single crop. This encourages farmers to concentrate on planning for the long term and to give more regard to the effects their actions will have on the environment. INM is dependent on a variety of aspects, such as the application of nutrients in an optimal manner and the conservation of those nutrients, as well as the dissemination of information on INM methods to farmers and researchers. Increasing the amount of nutrients that plants get may be accomplished via a variety of methods that are discussed in this book. Some of these methods include terracing, alley cropping, conservation tillage, intercropping, and crop rotation. In light of the fact that these technologies are discussed in other parts of this handbook, the emphasis of this section will be on INM in relation to the use of fertiliser in an appropriate manner. INM practices include new techniques that have been developed to improve nutrient uptake. These techniques include deep placement of fertilisers and the use of inhibitors or urea coatings (the use of area coating agent helps to retart the activity and growth of the bacteria responsible for denitrification). In addition to the standard selection and application of fertilisers, INM practices include these new techniques.

Notable Elements that Make up the INM Method are as follows

Testing methodologies to detect the availability of nutrients in plants and soils, as well as deficits in those nutrients. These include:

Visual signals may offer signs of particular nutrient deficits, which can be determined by plant symptom assessments. As an example, plants that are lacking in nitrogen seem to be stunted and pale in comparison to plants that are healthy.

Tissue analysis and soil testing — in cases when symptoms are not readily apparent, post-harvest tissue and soil samples may be analysed in a laboratory and compared with a reference sample taken from a plant that is in good condition.

- An in-depth analysis of the limitations and possibilities that are present in the existing soil fertility management techniques, as well as the ways in which these factors are connected to the nutrient diagnostic, such as the usage of fertilizers that are either inadequate or excessive.
- An analysis of the ways in which agricultural systems may be made more productive and sustainable. The appropriate proportion of nutrients required is determined by a variety of factors, including climate, soil type, crop type, agricultural techniques, and technological

advancements. Once these aspects have been comprehended, it will be possible to choose the most suitable INM technologies.

- Experimentation and development of INM technology driven by farming communities that are participatory. As a result of the need for technologies that are suitable for the local environment, it is necessary for farmers to participate in the testing and evaluation of any INM technology.

A Component-Based Approach to Integrated Nutrient Management

- On-site Resource Generation: The recycling of agricultural leftovers, animal manure, and other waste products is required for on-site resource creation.
- The mobilization of off-site nutrition resources necessitates the addition of chemical nutrients derived from outside sources. This is necessary in order to meet the requirements of the process.
- Integration of Resources: Chemical forms of nutrients and other management components that boost productivity need to be appropriately integrated with the resources that are responsible for the generation of energy and nutrients on-site.
- The integrated nutrition supply system requires the management of the agricultural system as a whole, which includes the use of chemicals to convert plant, animal, and poultry resources into food grains and other forms of food. This is referred to as "resource management."

Methods of Integrated Nutrient Management

Integrated Nutrient Management (INM) is a comprehensive approach to soil fertility management in agronomical crops that aims to optimize nutrient availability, enhance crop productivity, and promote sustainable agricultural practices. INM involves the judicious use of various nutrient sources and management practices, including chemical fertilizers, organic amendments, biofertilizers, green manures, crop rotations, and precision agriculture technologies. By integrating multiple nutrient sources and management strategies, INM seeks to improve soil health, reduce nutrient losses, minimize environmental impacts, and ensure long-term agricultural sustainability.

1. **Soil Testing and Nutrient Diagnosis:** One of the foundational principles of INM is soil testing and nutrient diagnosis. Soil tests provide valuable information about the nutrient status, pH levels, and organic matter content of the soil. By analysing soil samples, farmers can determine

nutrient deficiencies or imbalances and make informed decisions about fertilizer application.

2. **Balanced Fertilization:** INM advocates for balanced fertilization, which involves using a combination of chemical fertilizers, organic amendments, and biofertilizers to supply nutrients to crops. This approach ensures that crops receive a balanced diet of essential nutrients, promoting healthy growth and optimal yields.

3. **Organic Amendments:** Organic amendments such as compost, manure, and crop residues are essential components of INM. These materials improve soil structure, increase water retention, enhance microbial activity, and provide slow-release nutrients to crops. Incorporating organic matter into the soil helps build soil organic carbon, improve soil fertility, and sustain soil health over the long term.

4. **Biofertilizers:** Biofertilizers are microbial inoculants that enhance nutrient availability and uptake by plants. Common biofertilizers include nitrogen-fixing bacteria (e.g., rhizobium), phosphate-solubilizing bacteria, and mycorrhizal fungi. By forming symbiotic relationships with crop roots, these beneficial microorganisms help improve soil fertility, reduce the need for chemical fertilizers, and enhance crop yields.

5. **Green Manures and Cover Crops:** Green manures and cover crops are an integral part of INM systems. These crops, typically legumes such as clover, vetch, and alfalfa, are grown specifically to improve soil fertility and structure. When incorporated into the soil, green manures add organic matter, fix atmospheric nitrogen, suppress weeds, and enhance soil microbial activity.

6. **Crop Rotation:** Crop rotation is a key strategy in INM that involves alternating different crops in a sequence over time. Crop rotation helps break pest and disease cycles, improve soil structure, replenish soil nutrients, and reduce weed pressure. By diversifying cropping systems, farmers can optimize nutrient utilization, minimize nutrient losses, and maintain soil health.

7. **Nutrient Recycling:** Nutrient recycling is an essential component of INM, where crop residues, animal manure, and other organic materials are recycled back into the soil to replenish nutrients. Nutrient recycling helps close the nutrient loop, minimize nutrient losses, and improve nutrient use efficiency in agricultural systems.

8. **Precision Nutrient Management:** Precision agriculture technologies, such as soil sensors, remote sensing, and geographic information systems (GIS), play a crucial role in INM. These technologies enable farmers to apply fertilizers at variable rates based on soil nutrient levels, crop requirements, and spatial variability within fields. Precision nutrient management helps optimize fertilizer use efficiency, reduce nutrient losses, and improve crop yields.

9. **Nutrient-Use Efficient Varieties:** Selecting crop varieties that are efficient in nutrient uptake and utilization is an important strategy in INM. Nutrient-use efficient varieties have traits that enable them to thrive in low-nutrient conditions, utilize nutrients more efficiently, and produce higher yields with fewer inputs.

10. **Intercropping:** Intercropping involves growing two or more crops together in the same field at the same time. This practice promotes synergistic interactions between crops, enhances nutrient cycling, improves soil health, and increases overall productivity. Intercropping also helps reduce pest and disease pressure and provides greater resilience to environmental stresses.

11. **Agroforestry:** Agroforestry is a sustainable land-use system that integrates trees and shrubs with agricultural crops. In agroforestry systems, trees provide multiple benefits, including shade, windbreaks, erosion control, and nutrient cycling. Agroforestry enhances soil fertility by increasing organic matter content, improving nutrient cycling, and promoting biodiversity.

12. **Vermicomposting:** Vermicomposting is the process of using earthworms to decompose organic waste materials into nutrient-rich compost. Vermicompost is a valuable soil amendment that improves soil structure, enhances microbial activity, and provides a slow-release source of nutrients to crops.

13. **Biodynamic Farming:** Biodynamic farming is a holistic approach to agriculture that emphasizes the interconnectedness of soil, plants, animals, and the environment. Biodynamic practices, such as specific compost preparations and herbal treatments, are used to enhance soil fertility, promote plant health, and improve crop quality.

14. **Mulching:** Mulching involves covering the soil surface with organic or synthetic materials to conserve soil moisture, suppress weeds, and regulate soil temperature. Mulches also add organic matter to the soil as they decompose, improving soil fertility and structure over time.

15. **Integrated Pest Management (IPM):** Integrated Pest Management (IPM) is an ecologically-based approach to pest management that integrates multiple control strategies, including biological control, cultural practices, and chemical interventions. By reducing pest pressure, IPM helps minimize crop damage, improve plant health, and enhance nutrient uptake by crops.

Overall, Integrated Nutrient Management (INM) offers a holistic approach to optimizing soil fertility and enhancing crop productivity in agronomical crops. By integrating multiple nutrient sources and management practices, INM promotes sustainable agricultural practices, improves soil health, and ensures long-term agricultural sustainability.

Integrated Nutrient Management (INM) practices can be applied to a wide range of crops to optimize soil fertility and enhance productivity sustainably. Here are some examples of crops commonly grown with INM:

1. **Maize:** Maize is a staple food crop grown worldwide, and INM practices can significantly improve its productivity. Combining organic amendments such as compost or manure with balanced chemical fertilization can provide the necessary nutrients for maize growth. Additionally, intercropping maize with leguminous cover crops or incorporating crop residues as green manure can enhance soil fertility and reduce the dependency on external inputs.

2. **Rice:** Rice is another important crop that benefits from INM practices. Incorporating organic matter into rice paddies through the application of compost or green manure helps maintain soil fertility and improve water retention. Biofertilizers such as nitrogen-fixing bacteria can also be used to supplement nitrogen requirements, reducing the need for synthetic fertilizers. Furthermore, crop rotation with legumes or other non-rice crops can help break pest and disease cycles and improve soil health.

3. **Wheat:** Wheat is a major cereal crop grown in various agroecological zones, and INM practices can contribute to its sustainable production. Applying balanced fertilization with organic and inorganic sources can ensure optimal nutrient supply for wheat growth. Intercropping wheat with legumes or incorporating leguminous cover crops can enhance soil nitrogen levels and reduce fertilizer requirements. Precision nutrient management techniques can also be employed to adjust fertilizer application rates based on soil nutrient status and crop demand.

4. **Soybean:** Soybean is a leguminous crop that can fix atmospheric nitrogen with the help of symbiotic bacteria in its root nodules. However, supplementing nitrogen fixation with balanced fertilization and organic amendments can improve soybean yields and overall soil fertility. Intercropping soybean with cereals or using relay cropping systems can maximize land use efficiency and nutrient cycling. Additionally, incorporating soybean residues back into the soil as green manure can replenish nutrients and enhance soil organic matter content.

5. **Cotton:** Cotton is a cash crop grown for its fiber, and INM practices can improve its yield and quality while minimizing environmental impacts. Incorporating organic amendments such as compost or crop residues can improve soil structure and water retention, reducing irrigation requirements. Precision nutrient management techniques can optimize fertilizer application to meet cotton's nutrient demands at different growth stages. Additionally, intercropping cotton with legumes or cover crops can enhance soil fertility and suppress weeds.

Concerning the Limitations of Integrated Nutrient Management

- An insufficient amount of knowledge: Farmers usually do not possess the information that is required to apply fertilizers in a proportion that is balanced.

- Financial resources: Farmers, especially those living in rural regions, may not have access to loans. They are unable to acquire fertilizer and manure, both of which are necessary for INM, since they do not have sufficient funds.

- Degradation of Land: One of the most significant difficulties that the International Nature Conservation (INM) faces is the degradation of lands as a result of intense farming and over-exploitation brought on by the enormous strain brought on by the ever-increasing population.

- Variations in the Monsoon: The monsoon is important to the agricultural sector in India. The potential for water scarcity during drought-prone seasons is considered to be the most significant obstacle that stands in the way of the use of fertilizers. Erosion caused by water presents a significant threat to the fertility and production of soil during the monsoon season.

- Because the land in India is split into tiny holdings, the majority of farmers in the nation have small holdings. This is a limitation of small holdings. The usage of INM on a commercial basis is rendered difficult as a result of this.

- The use of biofertilizers is limited to certain crops and geographical locations, has a short shelf life, needs careful handling, and has other problems that make it difficult to use and distribute in an effective manner. These limitations make it difficult to employ biofertilizers.

Advantages

Integrated Nutrient Management (INM) offers numerous advantages in agricultural crop production, contributing to improved soil fertility, enhanced crop productivity, and sustainable agricultural practices. Some of the key advantages of INM include:

1. **Improved Soil Fertility:** INM combines the use of organic and inorganic nutrient sources to replenish soil nutrients and enhance soil fertility. Organic amendments such as compost and manure add organic matter and essential nutrients to the soil, improving soil structure, water retention, and nutrient availability. This leads to healthier soils that support vigorous plant growth and higher crop yields over the long term.
2. **Balanced Nutrition:** INM ensures that crops receive a balanced diet of essential nutrients, including nitrogen, phosphorus, potassium, and micronutrients. By combining various nutrient sources and management practices, INM provides crops with the nutrients they need at different growth stages, minimizing nutrient deficiencies and imbalances. Balanced nutrition promotes optimal plant growth, development, and yield potential.
3. **Reduced Environmental Impact:** INM practices promote environmentally sustainable agriculture by minimizing nutrient losses and reducing the risk of soil and water pollution. By using organic amendments and biofertilizers, INM helps reduce the dependency on synthetic fertilizers, which can leach into groundwater, contaminate surface water, and contribute to eutrophication of water bodies. Additionally, INM practices such as crop rotation, cover cropping, and reduced tillage help improve soil health, reduce erosion, and conserve natural resources.
4. **Cost Savings:** INM can lead to cost savings for farmers by reducing the need for expensive chemical fertilizers and minimizing input costs. Organic amendments such as compost and manure are often more affordable than synthetic fertilizers and can be produced on-farm or sourced locally. By using a combination of organic and inorganic nutrient sources, farmers can optimize nutrient use efficiency and maximize returns on investment while minimizing financial risks.

5. **Enhanced Crop Resilience:** INM practices help improve crop resilience to environmental stresses such as drought, heat, and pest infestations. Healthy soils with balanced nutrient levels and diverse microbial populations support strong root development, efficient nutrient uptake, and better tolerance to biotic and abiotic stresses. Additionally, INM promotes biodiversity and ecosystem resilience, reducing the risk of crop failure and promoting long-term agricultural sustainability.
6. **Long-Term Sustainability:** INM promotes sustainable agricultural practices that maintain or improve soil health, productivity, and resilience over the long term. By integrating multiple nutrient sources and management strategies, INM helps build soil organic matter, enhance soil structure, and promote nutrient cycling and conservation. Sustainable soil management practices benefit future generations by ensuring the continued productivity and viability of agricultural systems while preserving natural resources and protecting the environment.

Conclusion

In conclusion, Integrated Nutrient Management (INM) stands as a holistic and sustainable approach to optimizing soil fertility, enhancing crop productivity, and promoting environmentally friendly agricultural practices. By integrating various nutrient sources and management strategies, INM offers numerous advantages, including improved soil fertility, balanced nutrition for crops, reduced environmental impact, cost savings, enhanced crop resilience, and long-term sustainability. The adoption of INM practices is crucial for addressing the challenges of modern agriculture, including soil degradation, nutrient depletion, environmental pollution, and climate change. By prioritizing soil health and employing diverse nutrient management techniques, farmers can optimize yields, minimize input costs, and mitigate the negative impacts of agricultural activities on the environment. Moving forward, it is imperative to promote awareness and adoption of INM practices among farmers, policymakers, and stakeholders. Supportive policies, incentives, and extension services can facilitate the implementation of INM at the farm level, fostering a transition towards more sustainable and resilient agricultural systems. Additionally, continued research and innovation in soil science, agronomy, and agricultural technology will further advance our understanding of INM principles and practices, enabling farmers to achieve greater productivity, profitability, and environmental stewardship in the years to come.

References

Dwivedi, B. S., Singh, V. K., Meena, M. C., Dey, A., & Datta, S. P. (2016). Integrated nutrient management for enhancing nitrogen use efficiency. Indian J. Fertil, 12, 62-71.

Finck, A. (1998). Integrated nutrient management: an overview of principles, problems and possibilities. Annals of Arid Zone, 37, 1-24.

Gruhn, P., Goletti, F., & Yudelman, M. (2000). Integrated nutrient management, soil fertility, and sustainable agriculture: current issues and future challenges. Intl Food Policy Res Inst.

Jat, L. K., Singh, Y. V., Meena, S. K., Meena, S. K., Parihar, M., Jatav, H. S., & Meena, V. S. (2015). Does integrated nutrient management enhance agricultural productivity. J Pure Appl Microbiol, 9(2), 1211-1221.

Khan, M. U., Qasim, M., Khan, I. U., Qasim, M., & Khan, I. U. (2007). Effect of integrated nutrient management on crop yields in rice-wheat cropping system. Sarhad Journal of agriculture, 23(4), 1019.

Srivastava, A. K., & Ngullie, E. (2009). Integrated nutrient management: Theory and practice. Dynamic Soil, Dynamic Plant, 3(1), 1-30.

Zhang, F., Cui, Z., Chen, X., Ju, X., Shen, J., Chen, Q., & Jiang, R. (2012). Integrated nutrient management for food security and environmental quality in China. Advances in agronomy, 116, 1-40.

6

Integrated Nutrient Management in Horticultural Crops

Bijay Kumar Baidya[1], Pradipto Kumar Mukherjee[2]
Anamika Pandey[3] and Anushka Singh[4]

[1]Department of Agriculture and Allied Sciences, CV Raman Global University, Bhubaneswar, Odisha - 752054
[2]Palli Siksha Bhavana(Institute of Agriculture), Visva –Bharati, West Bengal
[3]Department of Horticulture (Fruit Science), Naini Agricultural Institute SHUATS, Prayagraj, Uttar Pradesh, India
[4]Department of Horticulture, K.P.H.E.I. Affiliated by Prof. Rajendra Singh (Rajju Bhaiya) University, Prayagraj

Abstract

Integrated Nutrient Management (INM) plays a pivotal role in optimizing soil fertility and enhancing crop productivity in horticultural crops. This abstract delves into the principles, benefits, and challenges associated with implementing INM practices in horticulture. INM integrates various nutrient sources and management strategies, including organic and inorganic fertilizers, biofertilizers, green manures, cover crops, and precision agriculture technologies. By combining these approaches, INM aims to maintain soil health, improve nutrient availability, and promote sustainable crop production systems in horticultural crops. While INM offers numerous advantages such as improved soil fertility, balanced nutrition, reduced environmental impact, and long-term sustainability, its implementation may face challenges such as resource constraints, technical know-how, and market access. Nevertheless, with proper planning, education, and support, INM has the potential to revolutionize nutrient management practices in horticultural crops, leading to increased yields, profitability, and environmental stewardship in the horticulture sector.

Keywords: *Fertilizers, soil, resources, education, inorganic*

Introduction

Fruits are an essential component of the human diet, and India holds the position of the second-largest fruit producer worldwide. The nation's agricultural sector produces banana, papaya, and mango, contributing approximately 13% of the global fruit production. Nevertheless, the present fruit output in India falls short of satisfying 46% of the total demand. Fruits provide a substantial amount of vitamins and minerals that are crucial for sustaining good health and immunity. In addition, they include oils, lipids, and proteins that are used in different metabolic processes.

The overutilization of synthetic fertilizers to enhance fruit yield has resulted in problems such as economic inefficiency, ecological degradation, and potential injury to both plants and consumers. This has resulted in significant issues. In order to meet the requirements of plants and enhance productivity, a significant quantity of nutrients must be supplied to the soil. Continual crop production may result in the gradual depletion of nutrient reserves in the soil, which ultimately leads to soil degradation. Due to the exorbitant expense of inorganic fertilizers, there has been a growing focus on using bio-fertilizers and organic matter with inorganic fertilizers. Farmers are now exploring the use of the Integrated Nutrient Management system for cultivating fruit crops. This system incorporates innovative techniques such as deep fertilizer application and urea coatings to enhance plant absorption of nutrients and minimize nutrient wastage. Ultimately, fruits are essential components of human meals, although their over use has resulted in economic inefficiencies, environmental harm, and land deterioration.

Nutrients Removal of Crops

Under optimal management circumstances, horticultural crops have the capacity to assimilate 500 to 1000 kg of NPK per hectare per year, or perhaps even higher amounts. Horticultural crops that need a lot of nutrients are known as heavy feeders. To achieve good yields, it is necessary to apply the right amount of nutrients in a balanced manner. The table provided in Ganeshamurthy et al. (2015) presents the estimated amounts of nutrients that are taken up by many significant horticulture crops. Horticultural crops exhibit significant variation in their nutrient removal patterns, which is contingent upon the desired yield. The analysis of nutrient removal patterns in various fruit crops indicates that the amount of potassium (K) removed is often 3-4 times more than the amount of nitrogen (N) and phosphorus (P) removed, in comparison to annual vegetable crops. These results provide the foundation for determining the fertilizer needs. In most circumstances, it is often seen that the amount of nutrients removed from the crop exceeds the amount of nutrients provided, resulting in the depletion of nutrients from the soil, a process known as nutrient mining. Manures and fertilizers only provide a portion of the depleting plant nutrients.

The Primary Goals of INM

The many goals of integrated plant nutrient management are as follows:

- To increase the amount of nutrients that are available in the soil throughout the growth season
- To reduce the amount of inorganic fertilizer that is required

In order to strike a balance between the crop's nutritional requirements and the available nutrients from all possible sources.

- To achieve the highest possible level of functioning of the soil biosphere in relation to a certain function
- Reduce the amount of nutrients that are lost beyond the root zone due to volatilization, denitrification, surface runoff, and leaching: this is the recommendation made by Sharma et al. (2018).

Components of INM

The formation of fruit relies heavily on inorganic nitrogen (INM), which is an essential component since it supplies plants with the nutrients they need. Straight fertilizers, complex fertilizers, and mixed fertilizers are the three types that may be taken into consideration while describing this substance. The inorganic chemicals that make up these fertilizers are responsible for supplying plants with one or more of the critical nutrients that they need.

As a result of the biological decomposition of organic matter, organic manures are produced. These manures may be obtained from either plant or animal waste. Both bulky organic manures, which have a low concentration of nutrients and must be supplied in large volumes, and concentrated organic manures, which contain much more nutrients than bulky organic manures, are examples of the types of manures that may be included in this category.

Compost may be made from crop residue, which is the material that is left over after harvesting crops and the leftovers of the agricultural sector. Keeping an adequate supply of nutrients is crucial, and crop rotation is a key strategy for doing so. This is accomplished by growing a variety of crops on the same piece of land. The soil is prevented from being depleted, and plants are able to absorb nutrients more effectively as a result. Preparations known as bio-fertilizers are preparations that enable agricultural plants to absorb nutrients in a more efficient manner. These bio-fertilizers include either active or dormant cells of effective strains of microorganisms. Bio-fertilizers fall into a variety of categories, including those that fix nitrogen and dissolve phosphate. It has been claimed that bio-fertilizers have a favourable effect in the growth of fruit crops such as mango and strawberry flowers. All things considered, INM is an extremely important factor in guaranteeing the well-being and expansion of fruit crops.

Importance of INM in Horticultural Crops

In spite of the fact that India's food production has greatly grown as a result of excellent nutrient management, it is not viable to meet all of the country's nutritional needs by the use of inorganic fertilizers. Plant performance, resource efficiency, and environmental protection are all areas that have been improved by the use of integrated nutrient management (INM), which was created to address these challenges. In order to achieve sustainable crop production and enhanced soil health, integrated nutrient management (INM) comprises the use of manures, chemical fertilizers, and biological agents. It is the most effective method for making more efficient use of the resources that are available and for decreasing the amount of money spent on crop production. Deficiencies in nitrogen, phosphorus, and sulphur are the primary variables that restrict yields in the soils of India. It is the goal of integrated nutrient management (INM) to keep the fertility of the soil at a level that is optimal for crop yield, therefore maximizing the advantages that may be obtained from all conceivable sources of plant nutrients, both organic and inorganic. It is vital to use this strategy in order to manage nutrient depletion and excess, and it is advantageous for marginal farmers who are unable to afford to feed all of their crop nutrients via the implementation of expensive chemical fertilizations. The transition from an ecosystem dominated by forests to an environment dominated by agriculture is the cause of the loss of organic matter in the soil and the deterioration of soil fertility.

Constraints

Integrated Nutrient Management (INM) is essential for improving the production of subtropical fruit crops in India because it enhances the characteristics of the soil and lessens the effect that excessive use of inorganic fertilizer has on the environment via its reduction of the environmental impact. Nevertheless, there are a number of obstacles that stand in the way of integrated nutrient management (INM). These include the slow decomposition of organic matter, the limited availability of farmyard manure (FYM), a lack of awareness of new research and technological innovations, resistance to modern agricultural practices, and the significance of crop rotation in ensuring that nutrients are always available.

As a result of the labour-intensive nature of composting and the management of bulky organic manures, farmers often choose synthetic fertilizers. The fertilizer prescription in INM does not take into account soil testing, which are necessary in order to apply the appropriate quantity of fertilizer to the various types of fruits and vegetables. In addition, farmers may not want to devote six to eight weeks of their time to the production of a crop that uses green manure since they do not get any financial compensation for their efforts. Furthermore, the advantages of green manure are not immediately obvious, which is another reason why farmers avoid utilizing it. To summarise, integrated nitrogen management (INM) is an essential method for enhancing the quality of soil and mitigating the adverse effects on the environment that are caused by the overuse of inorganic fertiliser. A significant number of farmers, on the other hand, are reluctant to embrace contemporary agricultural techniques because they are resistive to new ideas and because the production of green manure crops requires a significant amount of time and work.

Methods of INM in Horticultural Crops

Integrated Nutrient Management (INM) is a holistic approach to soil fertility management that involves the judicious use of various nutrient sources and management practices to optimize soil health and enhance crop productivity in horticultural crops. In horticulture, which encompasses the cultivation of fruits, vegetables, ornamental plants, and other specialty crops, effective nutrient management is essential for achieving high yields, quality produce, and economic sustainability while minimizing environmental impacts. INM integrates multiple methods to ensure balanced nutrient supply, promote soil fertility, and support sustainable crop production practices. In this comprehensive exploration, we will delve into 15 key methods of Integrated Nutrient Management in horticultural crops, highlighting their importance, implementation strategies, and benefits.

1. **Soil Testing and Nutrient Diagnosis:** Soil testing forms the foundation of Integrated Nutrient Management in horticultural crops. It involves the analysis of soil samples to assess nutrient levels, pH, organic matter content, and other soil properties. Soil tests provide valuable information that guides nutrient management decisions, allowing growers to apply fertilizers and soil amendments based on crop requirements and soil conditions. Soil testing helps identify nutrient deficiencies or imbalances, enabling targeted interventions to optimize soil fertility and promote healthy crop growth.

2. **Balanced Fertilization:** Balanced fertilization is a core principle of INM, emphasizing the use of both organic and inorganic fertilizers to supply essential nutrients to horticultural crops. By combining chemical fertilizers, organic amendments, and biofertilizers, growers can ensure a balanced nutrient supply that meets crop requirements throughout the growing season. Balanced fertilization helps prevent nutrient deficiencies or excesses, promotes optimal plant growth, and enhances crop yields and quality.

3. **Organic Amendments:** Organic amendments, such as compost, manure, and crop residues, are valuable sources of nutrients and organic matter in INM. These materials improve soil structure, increase water retention, enhance microbial activity, and supply essential nutrients to horticultural crops. Incorporating organic amendments into the soil helps build soil organic matter, promote nutrient cycling, and improve soil fertility over time. Organic amendments also contribute to long-term soil health and sustainability in horticultural cropping systems.

4. **Biofertilizers:** Biofertilizers contain beneficial microorganisms that enhance nutrient availability and uptake by plants. Common biofertilizers used in horticulture include nitrogen-fixing bacteria (e.g., Rhizobium), phosphate-solubilizing bacteria, and mycorrhizal fungi. By forming symbiotic relationships with plant roots, biofertilizers help improve soil fertility, reduce the need for chemical fertilizers, and enhance crop yields and quality. Biofertilizers also contribute to the biological diversity and ecological balance of soil ecosystems in horticultural crops.

5. **Cover Cropping:** Cover cropping is a sustainable practice that involves growing specific plant species, often legumes or grasses, during periods when the main crop is not growing. Cover crops help improve soil health, suppress weeds, prevent erosion, and promote nutrient cycling in horticultural cropping systems. Leguminous cover crops, such as clover or vetch, fix atmospheric nitrogen, providing a natural source of nitrogen for subsequent crops. Cover cropping also enhances soil organic matter content, improves soil structure, and supports beneficial soil microorganisms.

6. **Intercropping:** Intercropping is the practice of growing two or more crop species together in the same field. In INM, intercropping is used to maximize resource use efficiency, enhance nutrient cycling, and promote biodiversity in horticultural crops. Companion crops with complementary nutrient requirements and growth habits are selected

to optimize nutrient uptake and utilization. Intercropping also helps suppress weeds, reduce pest and disease pressure, and improve overall productivity in horticultural cropping systems.

7. **Precision Nutrient Management:** Precision agriculture technologies play a crucial role in INM by enabling growers to apply fertilizers and other inputs with precision and efficiency. Soil sensors, remote sensing, geographic information systems (GIS), and global positioning system (GPS) technology are used to monitor soil nutrient levels, crop performance, and environmental conditions in horticultural crops. Precision nutrient management allows growers to make data-driven decisions, optimize fertilizer applications, and minimize nutrient losses, leading to improved resource use efficiency and environmental sustainability.

8. **Nutrient Recycling:** Nutrient recycling is a key component of INM that involves the efficient use of organic materials, such as crop residues, animal manure, and food waste, to replenish soil nutrients in horticultural crops. By returning organic matter to the soil, nutrient recycling helps improve soil fertility, reduce the need for external inputs, and promote sustainable nutrient management practices. Composting, mulching, and vermicomposting are common methods of nutrient recycling used in horticultural cropping systems.

9. **Mulching:** Mulching is a practice that involves covering the soil surface with organic or synthetic materials to conserve moisture, suppress weeds, and regulate soil temperature in horticultural crops. Organic mulches, such as straw, hay, or compost, provide additional benefits by gradually decomposing and enriching the soil with organic matter and nutrients. Mulching helps improve soil fertility, reduce nutrient leaching, and enhance crop growth and productivity in horticultural cropping systems.

10. **Nutrient-Use Efficient Varieties:** Selecting crop varieties that are efficient in nutrient uptake and utilization is an important strategy in INM. Nutrient-use efficient varieties have traits that enable them to thrive in low-nutrient conditions, utilize nutrients more efficiently, and produce higher yields with fewer inputs. Breeding programs and genetic engineering techniques are used to develop nutrient-use efficient varieties of horticultural crops, contributing to improved resource use efficiency and sustainability.

11. **Drip Irrigation and Fertigation:** Drip irrigation is an efficient method of delivering water and nutrients directly to the root zone of horticultural

crops. Fertigation, a combination of irrigation and fertilization, involves applying fertilizers through the irrigation system, allowing for precise nutrient delivery and uptake by plants. Drip irrigation and fertigation systems help optimize nutrient use efficiency, minimize nutrient losses, and improve crop yields and quality in horticultural cropping systems.

12. **Nutrient Management Planning:** Nutrient management planning involves developing customized nutrient management plans based on soil test results, crop requirements, and environmental considerations in horticultural crops. Nutrient management plans specify the type, rate, timing, and placement of fertilizers and soil amendments to optimize nutrient supply and minimize environmental impacts. By following nutrient management plans, growers can make informed decisions, improve resource use efficiency, and ensure sustainable nutrient management practices in horticultural cropping systems.

13. **Soil Amendments:** Soil amendments, such as gypsum, lime, and elemental sulfur, are used to correct soil pH and improve nutrient availability in horticultural crops. Soil amendments help maintain optimal soil pH levels, which is essential for proper nutrient uptake and crop growth. They also enhance soil structure, water infiltration, and root development, leading to improved soil fertility and crop productivity in horticultural cropping systems.

14. **Controlled-Release Fertilizers:** Controlled-release fertilizers are designed to release nutrients slowly and steadily over time, providing a consistent supply of nutrients to horticultural crops. These fertilizers help reduce nutrient losses through leaching and volatilization, minimize nutrient runoff, and improve nutrient use efficiency. Controlled-release fertilizers are particularly beneficial in horticultural cropping systems where precise nutrient management is essential for maximizing yields and quality.

15. **Crop Residue Management:** Managing crop residues, such as stalks, stems, and leaves, is an important aspect of INM in horticultural crops. Crop residues contain valuable nutrients and organic matter that can be recycled back into the soil to replenish soil nutrients and improve soil fertility. Incorporating crop residues into the soil through tillage or mulching helps enhance soil organic matter content, promote nutrient cycling, and support healthy crop growth in horticultural cropping systems.

Advantages

Integrated Nutrient Management (INM) offers numerous advantages in horticultural crop production, contributing to improved soil fertility, enhanced crop productivity, and sustainable agricultural practices. Some of the key advantages of INM in horticulture crops include:

1. **Improved Soil Health:** INM practices focus on maintaining soil health by replenishing soil nutrients, enhancing soil structure, and promoting beneficial microbial activity. By incorporating organic amendments, biofertilizers, and cover crops, INM improves soil organic matter content, increases water retention, and enhances nutrient cycling, leading to healthier and more resilient soils in horticultural cropping systems.
2. **Balanced Nutrition:** INM ensures that horticultural crops receive a balanced diet of essential nutrients, including nitrogen, phosphorus, potassium, and micronutrients. By combining various nutrient sources and management practices, INM provides crops with the nutrients they need at different growth stages, minimizing nutrient deficiencies or imbalances and optimizing plant growth, development, and yield potential.
3. **Reduced Environmental Impact:** INM practices promote environmentally sustainable agriculture by minimizing nutrient losses, reducing the risk of soil and water pollution, and conserving natural resources. By using organic amendments, biofertilizers, and precision nutrient management techniques, INM helps reduce the dependency on synthetic fertilizers, which can leach into groundwater, contaminate surface water, and contribute to environmental degradation. Additionally, INM practices such as cover cropping and reduced tillage help improve soil health, reduce erosion, and mitigate climate change impacts in horticultural cropping systems.
4. **Cost Savings:** INM can lead to cost savings for horticultural growers by reducing the need for expensive chemical fertilizers and minimizing input costs. Organic amendments such as compost and manure are often more affordable than synthetic fertilizers and can be produced on-farm or sourced locally. By using a combination of organic and inorganic nutrient sources, growers can optimize nutrient use efficiency and maximize returns on investment while minimizing financial risks associated with nutrient management.

5. **Enhanced Crop Quality:** INM practices contribute to improved crop quality in horticultural crops by providing a balanced supply of nutrients, promoting optimal plant growth and development, and reducing the incidence of nutrient-related disorders. Balanced nutrition helps enhance fruit size, colour, flavour, and nutritional content, resulting in higher-quality produce that commands premium prices in the market.
6. **Resilience to Environmental Stresses:** INM practices help improve the resilience of horticultural crops to environmental stresses such as drought, heat, and pest infestations. Healthy soils with balanced nutrient levels and diverse microbial populations support strong root development, efficient nutrient uptake, and better tolerance to biotic and abiotic stresses. Additionally, INM promotes biodiversity and ecosystem resilience, reducing the risk of crop failure and promoting long-term agricultural sustainability.
7. **Compliance with Regulatory Requirements:** INM practices help horticultural growers comply with regulatory requirements related to nutrient management, environmental protection, and food safety. By adopting environmentally sustainable practices and minimizing nutrient losses, growers can reduce their environmental footprint and demonstrate their commitment to responsible stewardship of natural resources and ecosystems.

Conclusion

In conclusion, Integrated Nutrient Management (INM) stands as a cornerstone of sustainable horticultural crop production, offering a multifaceted approach to soil fertility management that balances environmental stewardship with economic viability. By integrating various nutrient sources and management practices, INM promotes soil health, enhances crop productivity, and fosters resilience to environmental stresses. The adoption of INM practices in horticulture holds significant promise for addressing key challenges facing modern agriculture, including soil degradation, nutrient depletion, and environmental pollution. Through the implementation of INM, horticultural growers can achieve improved yields, higher-quality produce, and cost savings while minimizing their environmental footprint and complying with regulatory requirements. Moving forward, continued research, education, and extension efforts are essential to further advance the adoption of INM practices and ensure their widespread implementation across horticultural cropping systems. By embracing INM principles, horticultural producers can contribute to a more sustainable and resilient food system that meets the needs of present and future generations.

References

Chadha, K. L., & Shikhamany, S. D. (1990). Integrated nutrient management in horticultural crops. Soil Fertility and Fertilizer Use, 4, 357.

Dolker, D., Bakshi, P., Wali, V. K., Dorjey, S., Kour, K., & Jasrotia, A. (2017). Integrated nutrient management in fruit production. International Journal of Current Microbiology and Applied Sciences, 6(7), 32-40.

El-Ramady, H. R. (2014). Integrated nutrient management and postharvest of crops. Sustainable Agriculture Reviews: Volume 13, 163-274.

Ganeshamurthy, A. N., Kalaivanan, D., Selvakumar, G., & Panneerselvam, P. (2015). Nutrient management in horticultural crops. Indian Journal of Fertilisers, 11(12), 30-42.

Ortega, R. (2013, October). Integrated nutrient management in conventional intensive horticulture production systems. In II International Symposium on Organic Matter Management and Compost Use in Horticulture 1076 (pp. 159-164).

Sharma, B. D., Jatav, M. K., Balai, R. C., & Meena, A. (2021). Integrated nutrient management for horticultural crops in arid region. In Dryland Horticulture (pp. 29-61). CRC Press.

7

Crop Selection and Breeding Method in Agricultural Crops

Parth K Rathod[1], Hrushik K Vadodaria[2], Kuldeep N Dudhatra[3] and Riddhi S Karmata[4]

[1-3]Department of Genetic and Plant Breeding, Anand Agricultural University Anand, Gujrat
[4]Department of Genetic and Plant Breeding, Navsari Agricultural University Navsari, Gujarat

Abstract

Crop selection and breeding methods play a crucial role in agricultural crop improvement, contributing to increased productivity, resilience, and nutritional quality. This abstract explores various crop selection and breeding methods employed in agricultural crops, highlighting their significance, approaches, and impacts. Traditional breeding methods, such as mass selection, pedigree selection, and recurrent selection, have been instrumental in developing improved crop varieties with desirable traits. Additionally, modern biotechnological tools, including marker-assisted selection, genetic engineering, and genomic selection, offer novel approaches to accelerate the breeding process and introduce precise genetic modifications. The integration of conventional and advanced breeding techniques has revolutionized crop improvement efforts, enabling breeders to address evolving challenges such as climate change, pest and disease resistance, and nutritional enhancement. By leveraging diverse breeding methods, agricultural scientists and breeders can develop resilient, high-yielding crop varieties tailored to meet the needs of farmers, consumers, and the environment in a rapidly changing world.

Keywords: *crop, breeding, modern, biotechnology, genetic, disease*

Introduction

Through the use of both art and science, crop breeding is the process of enhancing significant agricultural plants for the benefit of humanity. Crop breeders attempt to improve the productivity and nutritional value of our food, fiber, pasture, and industrial crops via their efforts. In order to meet the needs of a growing global population with rising dietary demands, crops are essential. Additionally, the activity of crop breeders contributes to the improvement of environmental protection.

Since the beginning of agriculture, farmers have been selecting plants for desirable characteristics such as bigger seeds, more flavourful fruits, and other desirable characteristics. This method is known as plant breeding. In today's world, the breeding of plants is a collaborative effort between scientists and farmers.

Different techniques are used by plant breeders, depending on the manner of reproduction of crops. This process is known as self-fertilization, and it occurs when pollen from one plant fertilizes the reproductive cells or ovules of another plant. There is a phenomenon known as cross-pollination, in which pollen from one plant can only fertilize another plant. For example, runners from strawberry plants are an example of asexual propagation, in which the new plant is genetically identical to its parent plant. Apomixis, also known as self-cloning, is a process in which seeds are created asexually and their offspring are genetically identical to their parents. When it comes to developing appropriate breeding and selection strategies, the mode of reproduction of a crop is the decisive element. This is because the genetic makeup of the crop is determined by the mode of reproduction. Additionally, it is necessary to have knowledge of the way of reproduction in order to artificially manipulate it in order to produce enhanced varieties. The only breeding and selection strategies that are appropriate for a crop are those that do not disrupt the crop's natural condition or guarantee that it will remain in that form. The imposition of self-fertilization on cross-pollinating crops results in a significant decrease in the performance of such crops for several reasons, including those listed above.

For the purpose of education, plant breeding is broken down into four distinct categories: line breeding, population breeding, hybrid breeding, and clone breeding. Line breeding refers to autogamous crops, population breeding refers to allogamous crops, hybrid breeding refers to primarily allogamous crops; clone breeding refers to vegetatively propagated crops.

Crop Selection in Crops

Crop selection is a critical process in agricultural crop improvement, involving the identification and selection of superior crop varieties with desirable traits for cultivation. Several methods are employed in crop selection, each tailored to specific breeding goals, environmental conditions, and market demands. Here are some of the key methods of crop selection:

1. **Phenotypic Selection:** Phenotypic selection involves the evaluation and selection of plants based on their observable traits or characteristics. Breeders assess various morphological, physiological, and agronomic traits such as yield, disease resistance, plant height, flowering time, and fruit quality. Superior individuals exhibiting desired traits are selected as parents for the next generation, contributing to the improvement of crop varieties.

2. **Performance Testing:** Performance testing involves evaluating the performance of different crop varieties under field conditions to assess their agronomic performance, yield potential, and adaptability to specific environments. Field trials are conducted across multiple locations and seasons to assess the stability and consistency of performance across diverse agro ecological conditions. Varieties with superior performance in field trials are selected for further evaluation or commercial release.

3. **Breeding Value Estimation:** Breeding value estimation utilizes statistical models to estimate the genetic merit or breeding value of individual plants based on their performance and pedigree information. By analysing data from field trials and pedigree records, breeders can estimate the genetic potential of plants for specific traits such as yield, disease resistance, and stress tolerance. Plants with high estimated breeding values are selected as parents for breeding programs to enhance the genetic potential of future generations.

4. **Marker-Assisted Selection (MAS):** Marker-assisted selection (MAS) is a molecular breeding approach that allows breeders to select plants with desirable traits based on molecular markers linked to specific genes of interest. Molecular markers are DNA sequences associated with traits of interest, such as disease resistance genes or quantitative trait loci (QTL) for yield or quality traits. By analysing molecular marker data, breeders can identify plants with desired traits at the molecular level, accelerating the selection process and improving breeding efficiency.

5. **Genomic Selection:** Genomic selection is a breeding approach that utilizes genome-wide marker information to predict the genetic merit of individual plants. By analysing the entire genome of a plant, breeders can

accurately estimate its breeding value for multiple traits, enabling more precise selection of superior genotypes. Genomic selection integrates genetic and phenotypic information to enhance breeding efficiency and accelerate genetic gain in crop improvement programs.

6. **Participatory Plant Breeding:** Participatory plant breeding involves collaboration between breeders, farmers, and other stakeholders in the selection and evaluation of crop varieties. Farmers actively participate in the selection process by evaluating and providing feedback on candidate varieties under their local growing conditions. Participatory approaches ensure that breeding programs address the needs and preferences of end-users, leading to the development of varieties that are well-suited to farmers' needs and preferences.

7. **Multi-Environment Trials (METs):** Multi-environment trials (METs) are conducted across multiple locations and seasons to evaluate the performance of crop varieties under different environmental conditions. METs provide valuable information on the genotype-by-environment interaction (GxE) and help breeders identify varieties with broad adaptation and stability across diverse agro ecological conditions. Varieties that perform consistently well across multiple environments are selected for further evaluation or release.

Breeding Methods

Autogamous Crops

Mass Selection

Mass selection refers to a breeding method where individuals with desirable traits are chosen for reproduction in order to improve the overall quality of a population. During mass selection, seeds are gathered from a population of individuals who exhibit desired traits, often ranging from a few dozen to a few hundred. The subsequent generation is then planted using a mixture of these collected seeds. This process, also known as phenotypic selection, relies on the physical appearance of each individual. Mass selection has been extensively used to enhance traditional "land" types, which are inherited from one generation of farmers to the next over extended periods. A long-standing alternative method, undoubtedly used for centuries, involves the eradication of unwanted individuals by eliminating them directly in their natural habitat. The outcomes remain same whether superior plants are preserved or inferior plants are eradicated: seeds from the superior plants are used as the planting material for the subsequent season. An updated method of mass selection involves the individual harvesting of the finest plants, followed by the cultivation and comparison of their offspring. The less affluent offspring are eliminated and the seeds of the remaining individuals are collected. It is important to highlight that the selection process now considers not only the physical characteristics of the parent plants but also the physical attributes and performance of their offspring. Progeny selection is generally more efficient than phenotypic selection for quantitative traits with modest heritability. It is important to mention that progeny testing requires an additional generation. Therefore, in order to obtain the same rate of gain per unit time, the gain per cycle of selection must be twice as much as that of basic phenotypic selection. Mass selection, whether accompanied by progeny test or not, is perhaps the most straightforward and cost-effective method among plant-breeding techniques. It is extensively used in the breeding of certain fodder species that do not have enough economic significance to warrant more thorough consideration.

Pure-line Selection

Pure-line selection refers to the process of selectively breeding individuals within a population to maintain and enhance desirable traits, resulting in a genetically uniform and stable line of organisms.

Pure-line selection typically consists of three distinct steps: (1) selecting numerous plants with superior characteristics from a genetically diverse population; (2) growing and evaluating the offspring of the selected plants through simple observation, often over several years; and (3) conducting extensive trials to determine the superiority of the remaining selections in terms of yield and other performance aspects, using meticulous measurements when observation alone is no longer sufficient. Any offspring that surpasses an existing variety is subsequently introduced as a novel "pure-line" variety. The effectiveness of this strategy in the early 1900s relied heavily on the presence of genetically diverse land kinds that were available for exploitation. They offered a plentiful supply of exceptional pure-line types, a few of which are still preserved in commercial kinds. Over the last several years, the pure-line strategy, as described earlier, has become less significant in the breeding of major farmed species. Nevertheless, this method continues to be extensively used for less significant species that have not undergone extensive selection. An ancient form of pure-line selection involves choosing single-chance variations, mutations, or "sports" from the original variety. A multitude of diverse kinds, exhibiting variations in traits like as colour, absence of thorns or barbs, reduced size, and resilience to diseases, have emerged via this process.

Hybridization

In the 20th century, deliberate crossbreeding of carefully chosen parents has become the prevailing method in breeding self-pollinating species. The purpose of hybridization is to merge advantageous genes present in two or more distinct kinds and generate offspring that are purebred and superior in several aspects compared to the original types. Genes, on the other hand, are consistently found with other genes in a group known as a genotype. The primary challenge faced by plant breeders is effectively controlling the vast quantities of genotypes that arise in the generations subsequent after hybridization. Hybridization, as shown by a hypothetical cross between wheat varieties that vary in just 21 genes, has the capacity to generate around 10,000,000,000 distinct genotypes in the second generation, showcasing its immense potential for generating variety. To accommodate every genotype in its predicted frequency, a population of such size would need almost 50,000,000 acres of land, using the typical spacing used by farmers. Although most of these second generation genotypes are hybrid (heterozygous) for one or more attributes, there is a statistical possibility of 2,097,152 distinct pure-breeding (homozygous) genotypes, each having the potential to be a new pure-line variety. The data demonstrate the need of using effective methods to manage hybrid populations, with the pedigree process being the most often used approach for this purpose.

Pedigree Breeding

Pedigree breeding begins by crossing two genotypes, each possessing one or more desired traits that are absent in the other. If the two original parents fail to possess all of the required traits, a third parent may be introduced by crossing it with one of the hybrid offspring from the first generation (F1). The pedigree approach involves the selection of better kinds in consecutive generations, while also keeping a record of parent-progeny ties.

The F2 generation, which is the offspring resulting from the mating of two F1 individuals, provides the first chance for selection in pedigree systems. In this period, the focus is on the eradication of people who possess unwanted main genes. In subsequent generations, the hybrid state transitions to pure breeding due to spontaneous self-pollination, causing families originating from various F2 plants to exhibit distinct characteristics. Typically, one or two exceptional plants are chosen from each excellent family throughout these generations. By the fifth generation, the state of being purebred (homozygosity) is widespread, and the focus moves mostly to selecting between families. The pedigree data is valuable for facilitating these eliminations. At this step, each chosen family is often harvested extensively to get higher quantities of seed required for evaluating families based on quantitative traits. This assessment is often conducted in plots cultivated under circumstances that closely mimic commercial planting practices. Once the number of families has been decreased to a reasonable size by visual selection, often by the F7 or F8 generation, a meticulous assessment for performance and quality begins. The comprehensive assessment of potential strains entails (1) meticulous observation, often conducted over many years and locales, to identify any previously unseen vulnerabilities; (2) rigorous evaluation of yield; and (3) thorough examination of quality. Plant breeders often conduct a five-year testing process at five specific sites to ensure the suitability of a new variety for commercial production.

The Bulk-population Approach

The bulk-population approach of breeding principally varies from the pedigree method in how it manages generations that come after hybridization. The F2 generation is planted at standard commercial planting densities on a spacious field. Upon reaching maturity, the crop is harvested in large quantities, and the seeds are used to establish the subsequent generation in a comparable plot. There is no documentation or record maintained about one's lineage or family history. During the phase of mass reproduction, natural selection seeks to remove plants with low survival value. There are two common methods of artificial selection: (1) eliminating plants that have undesired main genes, and (2) using mass approaches such as harvesting only partially developed seeds to choose early maturing plants, or using screens to select for larger seed size. Subsequently, individual plant choices are made and assessed using the same approach as in the pedigree breeding process. The primary benefit of the bulk population strategy is its cost-effectiveness in managing a substantial number of persons.

Frequently, an exceptional variety may be enhanced by introducing a certain desired trait that it now lacks. This may be achieved by first hybridizing a plant of the superior variety with a plant of the donor variety, which has the desired characteristic, then subsequently breeding the offspring with a plant that has the same genetic makeup as the superior parent. The term used to describe this phenomenon is backcrossing. After undergoing five or six rounds of backcrossing, the offspring will exhibit a hybrid trait that has been conveyed, while inheriting all other genes from the superior parent. By doing selfing on the last backcross generation, along with the process of selection, we may get offspring that exhibit pure breeding for the specific genes being transmitted. The backcross technique has many benefits, including its speed, the minimal number of plants needed, and the ability to accurately forecast the result. One significant drawback is that the technique reduces the likelihood of random pairings of genes, which might sometimes result in remarkable enhancements in performance.

Hybrid Cultivars

Hybridization and the generation of hybrid variants are distinct processes. The F1 hybrid resulting from crossings between distinct genotypes often exhibits significantly greater vigour compared to its parental genotypes. Heterosis, also known as hybrid vigour, may be expressed in several forms such as accelerated growth rate, enhanced uniformity, earlier blooming, and improved yield. The last is particularly significant in the field of agriculture.

Crops that are Fertilized by the Transfer of Pollen from one Plant to Another of the same Species.

Mass Selection

Mass selection refers to a breeding method where individuals with desirable traits are chosen for reproduction in order to improve the overall quality of a population. In cross-pollinated species, mass selection follows a similar process as in self-pollinated species. This involves selecting and harvesting a large number of plants that exhibit better traits, and then using their seeds to generate the following generation. The use of mass selection has shown significant efficacy in enhancing qualitative traits. Furthermore, when implemented across several generations, it has the potential to enhance quantitative traits, such as yield, despite the limited heritability of such traits. Mass selection is a widely used breeding technique for cross-pollinated species, particularly in economically less significant species.

Hybrid Cultivars

An exemplary demonstration of harnessing hybrid vigour via the use of F1 hybrid cultivars has been seen in the case of corn (maize). The process of developing a hybrid maize variety consists of three stages: (1) choosing better plants; (2) repeatedly self-pollinating them to create a set of inbred lines, each of which is distinct but genetically pure and very consistent; and (3) crossing the chosen inbred lines. During the process of inbreeding, the strength and vitality of the genetic lines falls significantly, often to less than half of that seen in types that are pollinated naturally in the field. The vigour is rejuvenated when two unrelated inbred lines are crossed, and in some instances, the F1 hybrids resulting from this crossbreeding exhibit much greater superiority compared to open-pollinated kinds. A significant consequence of the homozygosity of the inbred lines is that the resulting hybrid from crossing any two inbreds will consistently be identical. After identifying the most genetically diverse individuals that create the most favourable hybrids, it is possible to generate any required quantity of hybrid seeds.

Corn (maize) is pollinated by wind dispersal, where pollen is carried from the tassels to the styles (silks) that extend from the tops of the ears. Therefore, achieving regulated cross-pollination on a large scale may be done cost-effectively by planting two or three rows of the seed parent inbred alongside one row of the pollinator inbred, and removing the male flowers from the seed parent before they release pollen. Typically, the majority of hybrid corn is created by a process called "double crosses." This involves crossing four inbred lines in pairs (A × B and C × D), and then crossing the resulting two F1 hybrids again (A × B) × (C × D). The double-cross process offers the benefit of producing the commercial F1 seed using the highly prolific single cross A × B, rather than a low-yielding inbred, hence decreasing seed expenses. Recently, cytoplasmic male sterility has been used to avoid the need for removing the male flowers from the seed parent, resulting in cost savings in hybrid seed production. A significant portion of the hybrid vigour shown by F1 hybrid types diminishes in the subsequent generation. As a result, farmers do not utilize seed from hybrid types for planting, but instead they buy fresh seed each year from seed producers. The invention of hybrid corn (maize) has had a significant influence on expanding the number of food supplies accessible to the world's population, making it one of the most influential advancements in the biological sciences. The use of male sterility has enabled the development of hybrid variety in several crops, leading to significant success. It is quite probable that the adoption of hybrid varieties will further increase in the future.

Synthetic Cultivars

A synthetic variety is created by crossing many genotypes with known superior combining ability. These genotypes have been shown to provide better hybrid performance when crossed in various combinations. (In contrast, a mass selection variety consists of genotypes that have been bulked together without prior testing to evaluate their effectiveness in hybrid combinations.) Synthetic cultivars are renowned for their robust hybrid vigour and their capacity to provide viable seed for subsequent seasons. Due to these benefits, synthetic variations have gained popularity in the cultivation of several species, particularly fodder crops, when cost constraints prevent the creation or use of hybrid types.

Challenges in Breeding in Crops

People worldwide have a growing need for a wide range of nourishing meals. Ensuring global food security is a really challenging endeavour. The solution rests in using our croplands in the most sustainable manner possible. Agronomists strive to cultivate crop cultivars that exhibit enhanced productivity and nutritional content, even in the face of adverse environmental circumstances. When considering breeding programs, several aspects may be taken into account:

- Drought-tolerant and soil-conserving plants: These valuable resources are scarce and much sought after.
- Plants that preserve genetic variety: The greater the extent of our genetic diversity, the greater the ability of our crops to withstand future diseases or natural disasters. (For more insight into the significance of this, go to the Plant preservation section.)
- Plants with higher nutritional value are more efficient in using resources by providing more nutrients per calorie. (Further details about this topic may be found under the Nutritional quality section.)
- Plants with higher productivity on equivalent or smaller areas: It is imperative to restrict the further growth of agricultural fields in order to protect our forests and other natural places.
- Plant breeders also strive to enhance the adaptability of our crops to cope with the challenges posed by rising temperatures and unpredictable weather patterns.

Conclusion

Crop selection is a multifaceted process crucial to agricultural crop improvement, encompassing various methods tailored to specific breeding objectives and environmental conditions. Through phenotypic selection, performance testing, breeding value estimation, marker-assisted selection, genomic selection, participatory plant breeding, and multi-environment trials, plant breeders can identify and develop superior crop varieties with desirable traits such as high yield, disease resistance, and environmental adaptability. These selection methods, whether based on traditional breeding techniques or modern biotechnological tools, play a pivotal role in enhancing agricultural productivity, improving food security, and promoting sustainable agriculture. By leveraging the strengths of diverse selection methods and integrating them into comprehensive breeding programs, breeders can address the challenges of global agriculture and contribute to the development of resilient, high-performing crop varieties capable of meeting the needs of a growing population and a changing climate. Continued investment in crop selection research and breeding efforts is essential to ensure the long-term resilience and sustainability of agricultural systems worldwide.

References

Cabrera ,Bosquet, L., Crossa, J., von Zitzewitz, J., Serret, M. D., & Luis Araus, J. (2012). High-throughput phenotyping and genomic selection: The frontiers of crop breeding converge F. Journal of Integrative Plant Biology, 54(5), 312-320.

Crossa, J., Pérez-Rodríguez, P., Cuevas, J., Montesinos-López, O., Jarquín, D., De Los Campos, G., ... & Varshney, R. K. (2017). Genomic selection in plant breeding: methods, models, and perspectives. Trends in Plant Science, 22(11), 961-975.

Heffner, E. L., Sorrells, M. E., & Jannink, J. L. (2009). Genomic selection for crop improvement. Crop Science, 49(1), 1-12.

Kang, M. S., Subudhi, P. K., Baisakh, N., & Priyadarshan, P. M. (2007). Crop breeding methodologies: classic and modern. Breeding major food staples, 38-40.

Liu, J., Fernie, A. R., & Yan, J. (2021). Crop breeding–from experience-based selection to precision design. Journal of Plant Physiology, 256, 153313.

Moreno-Gonzalez, J., & Cubero, J. I. (1993). Selection strategies and choice of breeding methods. In Plant breeding: principles and prospects (pp. 281-313). Dordrecht: Springer Netherlands.

Tester, M., & Langridge, P. (2010). Breeding technologies to increase crop production in a changing world. Science, 327(5967), 818-822.

8

Integrated Pest Management in Agricultural Crops

Piyush Yadav[1], Hasim Kamal Mallick[2], P. Rajendra Gadge[3] and Ektedar[4]

[1]Agronomy, Integral Institute of Agricultural Science and Technology (IIAST), Integral University, Lucknow, Uttar Pradesh, India
[2]Institute of Agricultural Science, University of Calcutta, West Bengal
[3]Scientific Assistant, Department of Entomology, Raja Balwant Singh College of Agriculture, Agra, Uttar Pradesh, India
[4]Department of Plant protection, Chaudhary Charan Singh University Meerut, Uttar Pradesh, India

Abstract

Integrated Pest Management (IPM) is a holistic approach to pest control in agricultural crops that aims to minimize the use of chemical pesticides while effectively managing pest populations. This abstract explores the principles, strategies, and benefits of IPM in agricultural crop production. By integrating various pest management tactics such as cultural, biological, physical, and chemical control methods, IPM seeks to maintain pest populations at levels that do not cause economic damage while preserving ecosystem health and minimizing environmental risks. The adoption of IPM practices contributes to sustainable agriculture by reducing reliance on synthetic pesticides, minimizing pesticide residues in food and the environment, and promoting the conservation of natural enemies and biodiversity. Through the implementation of IPM, farmers can achieve effective pest control, optimize crop yields, and mitigate the adverse impacts of pest management practices on human health and the environment.

Keywords: *holistic, biological, agriculture, economics, risks, environmental*

Introduction

The development of Integrated Pest Management (IPM) was a response to the steadily increasing use of pesticides, which led to crises in pest control (outbreaks of secondary pests and pest resurgence following the development of pesticide resistance), as well as an increase in evidence and awareness of the full costs that the intensive use of pesticides imposes on human health and the environment. The term "Integrated Pest Management" (IPM) refers to the process of carefully considering all of the available methods of pest management and then subsequently incorporating suitable strategies that prevent the growth of pest populations. It integrates biological, chemical, physical, and crop-specific (cultural) management tactics and practices in order to cultivate healthy crops and avoid the use of pesticides. This helps to reduce or eliminate the hazards that pesticides cause to human health and the environment, which is essential for sustainable pest control. IPM is a dynamic process that makes use of an ecological systems approach and encourages the user or producer to examine and employ the complete range of optimal pest management choices available given economic, environmental, and social factors. IPM is accomplished via the use of an ecological systems approach. Integrated Pest Management (IPM) is founded on ecology, the idea of ecosystems, and the objective of maintaining ecosystem processes. Additionally, it stimulates the use of natural pest management techniques, which in turn promotes the establishment of a healthy crop while causing the least amount of damage to agroecosystems as feasible.

The Function of Integrated Pest Management in Environmentally Responsible Agriculture

- Implements environmentally responsible methods of pest management. Pollination is one of the ecosystem services that is protected by integrated pest management (IPM), which draws on ecosystem services such as insect predation. In addition to this, it helps to boost agricultural production and the availability of food by lowering the amount of crop losses that occur both before and after harvest.

- Reduces the amount of pesticide residues. IPM helps to ensure the safety of food and water by lowering the amount of pesticides that are used, which in turn lowers the amount of residues that are found in food, feed, and fiber, as well as in the environment.

- This improves the ecosystem's services. The objective of integrated pest management (IPM) is to preserve the equilibrium of the national agricultural environment. Conservation of the underlying natural resource base (such as soil, water, and biodiversity) and enhancement of ecosystem services (such as pollination, healthy soils, and variety of species) are both achieved by this practice.

- Leads to higher amounts of revenue. There is a reduction in the amount of pesticides used, which results in lower production costs. When crops are of higher quality and contain less residues, they are able to fetch higher prices in the market, which in turn contributes to enhanced profitability for farmers.
- Increases the farmer's level of expertise. IPM encourages stewardship on the part of farmers and develops their awareness of how ecosystems work in a manner that is tailored to their specific settings.

Integrating Culture into Pest Control: Through good farming methods, cultural control in integrated management lowers the number of pests. When crops are strong, cultural IPM methods work better. Because of this, regular EOSDA Crop Monitoring in integrated management helps deal with the issue quickly and lessen the bad effects that are about to happen. The following field management techniques are examples of IPM culture methods:

Treating the Soil: When the dirt is right for plants, they grow faster, and foods that are strong are less likely to get pests. In integrated pest management, checking the soil helps figure out if the field is good for growing a certain crop. If it isn't, chemicals can be added to make sure the plants grow well. Adding organic waste and mulch makes soil creatures more active and helps nutrients get into the soil. No-till farming helps keep the dirt from washing away, which makes farming more practical. But if you have to till the soil, you should do it in the fall so that the plants are not protected from natural enemies or bad weather.

Picking out the Right Plants: For crops to grow well, they need healthy plants and seeds, so it's important to pick growing materials that are free of pests and have strong roots. Certified seeds, eye inspection, and treating the seeds before planting help keep this from happening again. Planting resistant or tolerant types helps farmers keep their crops from going bad as part of an integrated pest control approach. As an example, a study shows that the hawthorn-carrot aphid can't hurt White Satin F1, Samba F1, Afro F1, Nipomo F1, and Yellowstone crops. Either by hand or with genetic engineering, stronger types are grown.

Rotation of Crops: Certain bug species can't live on non-host crop cycles. Rats and mice, for example, eat grain, and birds and worms hurt fruit. If the environment isn't right and there aren't any crops that need to be grown, pests will leave for places that are more profitable. This means that crop rotation can be used successfully as an integrated pest control method, among other things.

Planting Between Rows or Strip Cropping: When you intercrop or strip crop, different types of crops are planted between rows of host plants. This makes it take longer for pests to spread. This method is also used in the integrated pest management system. On the other hand, when plants of the same crop or family grow together, outbreaks happen more often. So, pests that used to only attack cabbage may now attack mustard, broccoli, and other brassicas. On the other hand, potato bugs can hurt both growing potatoes and tomatoes.

How to use Trap Plants: For IPM intercropping, you could also put trap plants in patches. This method of integrated pest management suggests using chemicals or mechanical methods to get rid of pests that have already been attracted to certain plants. One food that can be used to catch Japanese bugs is soybeans. Maggots that eat cabbage roots also like radishes.

Pick which Dates to Plant: When it comes to integrated pest management, the best times to plant or sow make the crops that are least likely to be attacked by pests or that are already strong enough to handle outbreaks. For example, if the earth is warm enough, it is best to plant squash early so that it can be ready to eat before pickleworm comes back from the south. At the same time, planting too early may cause root rot because the earth is too wet after winter.

Integrated Mechanical and Physical Pest Management: Integrated mechanical and physical controls mean getting rid of or killing pests with the right tools, by hand, or by making it impossible for them to get to the plants. As part of mechanical IPM control, traps and warming or burning the soil are used. Physical means of IPM include things like barriers. Let us check them out more.

Hand Picking: Manually getting rid of or picking out bugs takes a lot of time and work, but it is a common method used in integrated management and organic farming. Adult bugs or their eggs and larvae are hand-collected and thrown away.

Traps: One popular mechanical IPM way for getting rid of dangerous organisms is to use traps. There are different kinds of electric or mechanical traps that use light or fire to attract pests, air force to catch them, or electricity or sound to scare them away.

Putting Barriers: While making screens for birds and bugs or walls around fields to keep animals out can be helpful, there are times when they are not necessary. For instance, kangaroos in Australia can jump three meters high, so even the tallest fence would not stop them. There will be no point in putting up barriers; the case will need a different combined management system.

Cutting Back and Raking: Integrated management practices allow cutting off infected parts of plants when the damage can't be fixed and there is no way to treat the crop disease. Raking helps get rid of pests manually or by bringing them to the ground's surface, where predators (like birds) can eat them.

Taking Care of Irrigation: The right way to handle watering gives plants what they need to grow in a healthy way and keeps pests away. Integrated management systems can use different methods to apply pesticides, such as sprays and sprinklers on the leaves or drip watering. There are, however, both good and bad effects of watering control on IPM.

Treating with Heat or Steam: Using high temperatures to heat or steam grounds is an effective way to get rid of pests, their eggs and larvae, pathogens, and plant seeds. Heating and steaming can also be used to clean waste, organic matter, and farming tools. In this type of mechanical integrated pest control, fuels are burned to make water steam that treats the earth. Pasteurization (160–182°F) or cleaning (212°F) for 30 minutes can be used to heat the soil.

Biocontrol as Part of Integrated Pest Management: The idea behind this combined management method is that bugs can be killed by predators, parasitoids, viruses, and other biological control agents, also known as hostile organisms. The goal of biological control in IPM is to keep environments as balanced as possible by acting like nature does. In integrated management, biocontrol works like natural processes, but natural control happens without any help from people.

How to Use Predators: Predators eat their food that hurts plants. For example, ladybugs cut down on the number of aphids. For combined management, the numbers of predators are increased in their natural environment or brought in from other areas. But when raising the amount of enemies in integrated pest control, there are a few things to keep in mind:

"Foreign" predators might not be able to do the job; removing some species might lead to new pest invasions; animals that are brought into a new area may become pests over time if there are no natural enemies to keep their numbers in check.

When rabbits were brought into Australia, it was a well-known case of a food chain mistake in integrated management. As time went on, their numbers became a real problem for farms, just like native rabbits or dingoes. The cane toad is another example of how integrated biological control doesn't work in this situation because it wouldn't hunt the target species and turned into a pest.

How Parasitoids Are Used: Parasitoids grow on or inside their hosts and kill them when they're fully grown. The most common parasitoids are wasps and flies. When using this method of combined pest control, it is important to keep in mind that hyperparasitoids can also fight parasitoids. A study found that 20% of the cereal aphids in Canadian wheat areas were hyperparasitic.

How Pathogens Are Used: Pathogenic microorganisms are viruses, bacteria, and fungi that attack pests and make them sick, which lowers the number of pests. They are also used in the integrated pest control system. For example, the number of wild rabbits dropped by a lot after they got the myxomatosis virus from mosquitoes. A flea-borne virus was used to carry out the plan in places where mosquitoes were not common.

Chemical Methods: Integrated Pest Management (IPM) is an ecological approach to controlling pests that combines various management strategies and practices to grow healthy crops and minimize the use of pesticides. Chemical methods are one component of IPM and are used judiciously to manage pest populations. Here's how chemical methods fit into IPM and the principles guiding their use in agricultural crops:

Role of Chemical Methods in IPM

1. **Last Resort Strategy:** Chemical methods are often considered a last resort within IPM. This means they are used only when other methods (cultural, biological, mechanical, and physical) have failed to keep pest populations below economically damaging levels.

2. **Selective Pesticides:** When chemical control is necessary, IPM emphasizes the use of selective pesticides that target specific pests while minimizing harm to beneficial organisms, non-target species, and the environment.

3. **Threshold-Based Applications:** Chemical treatments are applied based on pest population thresholds. These thresholds are determined by monitoring pest levels and assessing the potential economic damage. Treatment is only carried out when pest populations exceed these thresholds.

4. **Resistance Management:** IPM strategies include rotating pesticides with different modes of action to prevent or delay the development of pest resistance. This ensures that pesticides remain effective over time.

Chemical Methods Used in IPM

1. **Insecticides:** These are used to control insect pests. In IPM, insecticides are chosen based on their effectiveness against the target pest and their environmental impact. Examples include neonicotinoids, pyrethroids, and insect growth regulators.
2. **Herbicides:** Used to manage weed populations. Selective herbicides target specific weeds without affecting the crop. Herbicides can also be used in combination with crop rotation and other cultural practices to manage weed resistance.
3. **Fungicides:** Applied to control fungal diseases. IPM programs advocate for the use of fungicides with different modes of action to prevent resistance and encourage the use of fungicides that are less harmful to non-target organisms.
4. **Rodenticides:** Used to control rodent populations. These are used strategically in combination with habitat modification and exclusion techniques to minimize their impact on non-target species.

Best Practices for Using Chemical Methods in IPM

1. **Correct Identification:** Proper identification of pests is crucial to selecting the appropriate pesticide and avoiding unnecessary applications.
2. **Targeted Application:** Applying pesticides in a targeted manner reduces the amount needed and minimizes exposure to non-target areas and organisms.
3. **Precision Agriculture:** Utilizing technologies such as GPS-guided equipment and drones to apply chemicals precisely and efficiently.
4. **Compliance with Regulations:** Following local and international regulations regarding pesticide use to ensure safety and environmental protection.

Legal Control

Legal methods in Integrated Pest Management (IPM) refer to the regulations, policies, and enforcement mechanisms established by governments and international bodies to ensure sustainable and safe pest management practices in agriculture. These legal frameworks help in minimizing the risks associated with pesticide use, protecting human health, and preserving the environment. Here's how legal methods are integrated into IPM for agricultural crops:

Regulatory Frameworks

1. **Pesticide Registration and Approval:** Governments require that pesticides be registered and approved before they can be marketed and used. This involves rigorous testing for efficacy, safety, and environmental impact. For example, agencies like the Environmental Protection Agency (EPA) in the United States, the European Food Safety Authority (EFSA) in the EU, and similar bodies in other countries are responsible for this process.
2. **Maximum Residue Limits (MRLs):** Legal limits are set for the amount of pesticide residue that can be present on food crops. These limits ensure that the levels of pesticides on crops are safe for human consumption. MRLs are established based on scientific assessments and are enforced through regular monitoring and inspections.
3. **Labelling Requirements:** Regulations mandate that pesticide products carry labels with detailed information on their proper use, dosage, safety precautions, and potential risks. This helps ensure that users apply pesticides correctly and safely.

Enforcement and Compliance

1. **Inspection and Monitoring:** Regulatory agencies conduct inspections and monitoring of agricultural practices to ensure compliance with pesticide laws. This includes checking for proper usage, storage, and disposal of pesticides.
2. **Penalties and Sanctions:** Legal frameworks include penalties and sanctions for non-compliance. Farmers and businesses that violate pesticide regulations can face fines, restrictions, or other legal actions.

Supportive Policies

1. **Subsidies and Incentives:** Governments may provide subsidies or financial incentives to farmers who adopt IPM practices. This can include funding for training programs, research, and the adoption of non-chemical pest control methods.
2. **Research and Development:** Legal frameworks often support research and development of new pest control methods and technologies. This includes funding for public and private research institutions to develop safer and more effective pest management solutions.
3. **Education and Training:** Policies may mandate education and training programs for farmers and agricultural workers on IPM practices and safe

pesticide use. These programs aim to increase awareness and knowledge about sustainable pest management.

Conclusion

In conclusion, legal methods play a critical role in the effective implementation of Integrated Pest Management (IPM) in agricultural crops. Through rigorous regulatory frameworks, enforcement mechanisms, and supportive policies, these legal measures ensure that pest control practices are safe, sustainable, and environmentally responsible. By mandating the registration and approval of pesticides, setting maximum residue limits, and requiring proper labelling, regulatory bodies protect public health and the environment. Additionally, international agreements and standards harmonize efforts across borders, promoting global food safety. The integration of legal methods into IPM not only enhances compliance and accountability but also encourages the adoption of innovative and less harmful pest management strategies, ultimately contributing to the sustainability of agricultural practices and the protection of ecosystems.

References

Anderson, J. A., Ellsworth, P. C., Faria, J. C., Head, G. P., Owen, M. D., Pilcher, C. D., & Meissle, M. (2019). Genetically engineered crops: importance of diversified integrated pest management for agricultural sustainability. Frontiers in bioengineering and biotechnology, 7, 24.

Barzman, M., Bàrberi, P., Birch, A. N. E., Boonekamp, P., Dachbrodt-Saaydeh, S., Graf, B., & Sattin, M. (2015). Eight principles of integrated pest management. Agronomy for sustainable development, 35, 1199-1215.

Dhawan, A. K., Singh, B., Bhullar, M. B., & Arora, R. (2013). Integrated pest management. Scientific publishers.

Peshin, R., & Zhang, W. (2014). Integrated pest management and pesticide use (pp. 1-46). Springer Netherlands.

Smith, R. F., Apple, J. L., & Bottrell, D. G. (1976). The origins of integrated pest management concepts for agricultural crops. Integrated pest management, 1-16.

Trivedi, T. P., & Ahuja, D. B. (2011). Integrated pest management: approaches and implementation. Indian Journal of Agricultural Sciences, 81(11), 981-993.

9

Integrated Pest Management in Horticultural Crops

Pavithra A[1], Pramod Kumar M[2], Adama Thanuja[3] and Hariharan S[4]

[1,2]College of Agriculture, Rajendranagar, Professor Jayashankar Telangana State Agricultural University, Rajendranagar, Hyderabad
[3,4]Division of Entomology, ICAR-IARI, Mega University, Hyderabad Hub

Abstract

Integrated Pest Management (IPM) in horticultural crops is an ecologically-based strategy that combines various control methods to manage pest populations in an economically viable and environmentally sustainable manner. This approach integrates cultural, biological, mechanical, and chemical methods, with an emphasis on minimizing the use of chemical pesticides to reduce adverse effects on human health and the environment. IPM relies on regular monitoring and accurate identification of pests to implement timely and targeted interventions. Legal frameworks and regulations further support IPM by enforcing safe pesticide use and promoting research and education on sustainable practices. The adoption of IPM in horticulture not only enhances crop health and yield but also supports biodiversity and ecosystem services, making it a vital component of modern agricultural practices.

Keywords: *IPM, integrates, legal, safe, ecosystems*

Introduction

Integrated Pest Management (IPM) represents a holistic approach to managing pest populations in horticultural crops, aiming to minimize the use of chemical pesticides while ensuring sustainable and economically viable crop production. IPM integrates multiple strategies—cultural, biological, mechanical, and chemical—within a coordinated framework, emphasizing ecological balance and long-term pest suppression. This article provides a comprehensive overview of IPM principles, methodologies, and applications in horticultural crops, addressing the ecological, economic, and regulatory dimensions of this essential agricultural practice.

Principles of IPM

Ecological Approach

IPM is fundamentally an ecological approach that considers the complex interactions between pests, crops, and the environment. By understanding these interactions, IPM seeks to manipulate ecosystems to favour natural pest control mechanisms and reduce the need for synthetic chemical inputs.

Multiple Control Strategies

IPM employs a combination of control strategies

- **Cultural Controls:** Practices such as crop rotation, intercropping, and selecting pest-resistant varieties to create unfavourable conditions for pests.
- **Biological Controls:** Utilizing natural predators, parasites, and pathogens to control pest populations.
- **Mechanical and Physical Controls:** Methods like traps, barriers, and manual removal to reduce pest numbers.
- **Chemical Controls:** Judicious use of pesticides when necessary, chosen for their specificity and minimal non-target effects.

Monitoring and Decision-Making

Regular monitoring of pest populations and environmental conditions is a cornerstone of IPM. This involves:

- **Scouting:** Systematic sampling and observation of crops to detect pest presence and population levels.
- **Thresholds:** Establishing action thresholds, the pest population levels at which control measures must be taken to prevent economic damage.
- **Decision Support Systems:** Utilizing models and tools to predict pest outbreaks and guide timely interventions.

Resistance Management

To delay the development of pest resistance, IPM includes strategies such as rotating pesticides with different modes of action and integrating non-chemical methods.

IPM Strategies in Horticultural Crops

Cultural Controls

Crop Rotation: Rotating crops disrupts pest life cycles, reducing their populations. For instance, alternating between leafy greens and root vegetables can minimize soil-borne pests that specialize in one type of plant.

Intercropping: Planting different crops together can confuse pests and reduce their ability to locate their preferred host plants. This diversity can also enhance the presence of beneficial organisms.

Sanitation: Removing crop residues and weeds that can harbour pests helps prevent future infestations. Practices such as cleaning equipment and using certified disease-free seeds also contribute to sanitation.

Biological Controls

Natural Enemies: Introducing or conserving natural predators, such as lady beetles for aphid control, can significantly reduce pest populations. These biological control agents often provide self-sustaining pest suppression.

Parasitic Wasps: Parasitic wasps lay eggs in or on pest insects, eventually killing them. This method is particularly effective against caterpillar pests in many horticultural crops.

Pathogens: Using microbial pesticides, such as Bacillus thuringiensis (Bt), targets specific pest larvae without affecting non-target organisms. This approach is widely used in organic and conventional horticulture.

Mechanical and Physical Controls

Traps: Pheromone traps and sticky traps are used to monitor and directly reduce pest populations. These traps are especially useful for early detection and management of flying insects like moths and flies.

Barriers: Physical barriers, such as row covers and insect netting, protect crops from pest access. These methods are effective in preventing infestations of pests like aphids and whiteflies.

Manual Removal: For small-scale operations, manual removal of pests, such as handpicking caterpillars or using water sprays to dislodge aphids, can be effective and environmentally friendly.

Chemical Controls

Selective Pesticides: When chemical intervention is necessary, selective pesticides that target specific pests while sparing beneficial organisms are preferred. This minimizes ecological disruption and reduces the likelihood of pest resistance.

Reduced-Risk Pesticides: Newer, reduced-risk pesticides are designed to be safer for non-target species and the environment. These include biopesticides derived from natural materials and synthetic chemicals with low toxicity profiles.

Integrated Applications: Combining chemical treatments with other IPM strategies ensures that chemicals are used as a last resort, in conjunction with biological and cultural methods, to maintain pest populations below economic thresholds.

Implementation of IPM

Steps in IPM Implementation

1. **Pest Identification and Monitoring:** Accurate pest identification is critical for effective management. Regular monitoring through field scouting, trapping, and remote sensing helps track pest populations and environmental conditions.

2. **Establishing Action Thresholds:** Determining action thresholds for each pest helps avoid unnecessary interventions. These thresholds are based on economic, agronomic, and environmental considerations.

3. **Choosing Appropriate Control Methods:** Selecting the most effective and least disruptive control methods involves considering the pest's biology, crop stage, and environmental impact. Integrated strategies are prioritized to enhance sustainability.

4. **Implementing Controls:** Timely and precise application of control measures, whether cultural, biological, mechanical, or chemical, ensures effective pest management. Record-keeping and continuous evaluation help refine IPM practices.

Case Studies in Horticultural IPM

Tomato Production: Tomato crops are susceptible to pests like aphids, whiteflies, and tomato hornworms. An IPM program for tomatoes might include:

- Crop rotation and resistant varieties to reduce pest pressure.
- Introduction of parasitic wasps and predatory beetles.

- Use of pheromone traps for monitoring and controlling moth populations.
- Selective application of insecticides when pest thresholds are exceeded.

Apple Orchards

Apple orchards face challenges from pests such as codling moths and apple maggots. An effective IPM approach includes:

- Mating disruption techniques using pheromone dispensers.
- Conservation of natural enemies like parasitic wasps and predatory mites.
- Regular monitoring with traps and visual inspections.
- Application of targeted insecticides during critical periods.

Integrated Pest Management (IPM) in Mango Cultivation

Mango (Mangifera indica), known as the "king of fruits," is a significant tropical fruit crop valued for its delicious taste, nutritional benefits, and economic importance. However, mango cultivation faces various challenges from pest infestations that can significantly affect yield and fruit quality. Integrated Pest Management (IPM) in mango cultivation offers a sustainable approach to managing these pests by combining multiple strategies—cultural, biological, mechanical, and chemical—while minimizing environmental impact and promoting ecological balance.

Key Pests in Mango Cultivation

1. Mango Hoppers (Idioscopus spp.)

- **Damage:** Suck sap from tender parts of the tree, leading to wilting and reduced fruit set.
- **Symptoms:** Leaf curling, sooty mold development due to honeydew secretion.

2. Fruit Flies (Bactrocera spp.)

- **Damage:** Lay eggs in the fruit, leading to larval feeding and fruit rot.
- **Symptoms:** Premature fruit drop, maggot-infested fruits.

3. Mealybugs (Drosicha mangiferae and Rastrococcus spp.)

- **Damage:** Suck sap from various parts of the plant, weakening the tree and causing sooty mold.

- **Symptoms:** White cottony masses on plant parts, black sooty mold on leaves.

4. Stem Borers (Batocera rufomaculata)

- **Damage:** Larvae bore into stems and branches, causing dieback and structural damage.
- **Symptoms:** Sawdust-like frass near entry holes, branch wilting.

5. Anthracnose (Colletotrichum gloeosporioides)

- **Damage:** Fungal disease affecting leaves, flowers, and fruits, causing black spots and lesions.
- Symptoms: Dark, sunken lesions on fruit and leaves, leaf blight.

Principles of IPM in Mango

1. Monitoring and Identification

- **Scouting:** Regular field scouting to identify pest presence and population levels.
- **Traps:** Use pheromone and sticky traps for monitoring pests like fruit flies and mango hoppers.
- **Record-Keeping:** Maintain detailed records of pest populations and interventions.

2. Preventive Cultural Practices

- **Sanitation:** Remove and destroy infested fruits and plant debris to reduce pest breeding sites.
- **Pruning:** Regular pruning to improve air circulation and reduce habitat for pests.
- **Resistant Varieties:** Cultivate pest-resistant mango varieties to reduce susceptibility.

3. Biological Controls

- **Natural Enemies:** Conservation and augmentation of natural predators like lady beetles, parasitic wasps, and predatory mites.
- **Biopesticides:** Use microbial pesticides such as Bacillus thuringiensis (Bt) and entomopathogenic fungi to target specific pests.

4. Mechanical and Physical Controls

- **Traps and Barriers:** Use of pheromone traps for fruit flies and sticky bands on tree trunks to control mealybugs.
- **Manual Removal:** Handpicking and destroying visible pests like stem borers and their larvae.

5. Chemical Controls

- **Selective Pesticides:** Application of selective insecticides and fungicides when pest populations exceed economic thresholds.
- **Reduced-Risk Chemicals:** Use of reduced-risk pesticides that are less harmful to non-target organisms and the environment.
- **Integrated Applications:** Combine chemical treatments with other IPM strategies to reduce reliance on pesticides.

Integrated Pest Management (IPM) in Grapes

Key Pests and Diseases in Grapes

1. Grape Berry Moth (Paralobesia viteana)

- **Damage:** Larvae feed on grape berries, causing direct damage and making them susceptible to fungal infections.
- **Symptoms:** Webbing and frass in clusters, premature fruit drop.

2. Grape Phylloxera (Daktulosphaira vitifoliae)

- **Damage:** Root-feeding aphids cause root galls, leading to weakened vines and reduced yield.
- Symptoms: Swellings or galls on roots, yellowing leaves, stunted growth.

3. Powdery Mildew (Erysiphe necator)

- **Damage**: Fungal disease affecting leaves, stems, and berries, reducing photosynthesis and fruit quality.
- **Symptoms:** White, powdery growth on surfaces, distorted leaves, and berries.

4. Downy Mildew (Plasmopara viticola)

- **Damage:** Fungal disease that affects leaves, shoots, and clusters, causing defoliation and yield loss.

- **Symptoms:** Yellow oil spots on leaves, white downy growth on the underside of leaves.

5. Botrytis Bunch Rot (Botrytis cinerea)

- **Damage:** Fungus infects grape clusters, leading to rot and significant yield loss.
- **Symptoms:** Brown, soft, and watery berries, grey mold on clusters.

IPM Strategies for Major Grape Pests and Diseases

Grape Berry Moth

1. Cultural Practices

- Remove wild grapevines near vineyards to reduce alternate hosts.
- Timely harvesting to minimize exposure of ripe berries to moth infestation.

2. Biological Control

- Release of parasitic wasps like Trichogramma spp. to target grape berry moth eggs.
- Conservation of native predators and parasitoids.

3. Mechanical Control

- Use of pheromone traps to monitor and mass trap male moths, reducing mating success.
- Installation of pheromone dispensers for mating disruption.

4. Chemical Control

- Application of insecticides targeting larvae, timed according to moth activity and egg hatch.

Grape Phylloxera

1. Cultural Practices

- Use of resistant rootstocks to prevent root infestation.
- Maintain vine health through proper nutrition and irrigation to enhance resistance.

2. Biological Control

- Research on biocontrol agents like nematodes that can target phylloxera.

3. Mechanical Control

- Regular monitoring of roots for early detection of galls and damage.

4. Chemical Control

- Soil treatments with systemic insecticides, applied with caution to minimize impact on soil health.

Powdery Mildew

1. Cultural Practices

- Prune and train vines to improve air circulation and reduce humidity.
- Avoid excessive nitrogen fertilization, which can increase susceptibility.

2. Biological Control

- Application of biocontrol agents like Ampelomyces quisqualis, a fungal hyperparasite of powdery mildew.
- Use of sulphur-based biopesticides.

3. Mechanical Control

- Removal of infected plant parts to reduce inoculum sources.

4. Chemical Control

- Application of fungicides like potassium bicarbonate and myclobutanil, timed according to disease development and weather conditions.

Downy Mildew

1. Cultural Practices

- Ensure proper vineyard drainage to avoid waterlogged conditions that favor the disease.
- Use of resistant grape varieties.

2. Biological Control

- Application of biocontrol agents like Trichoderma spp. to reduce pathogen load.
- Use of copper-based biopesticides.

3. Mechanical Control

- Removal and destruction of infected plant material.

4. Chemical Control

- Application of systemic and contact fungicides, such as metalaxyl and mancozeb, during periods of high risk.

Botrytis Bunch Rot

1. Cultural Practices

- Improve air circulation through proper canopy management and spacing.
- Remove infected plant material and mummified berries.

2. Biological Control

- Use of biocontrol agents like Bacillus subtilis and Trichoderma spp. to suppress Botrytis.
- Application of plant defense stimulants.

3. Mechanical Control

- Use of physical barriers such as paper bags to protect clusters during ripening.

4. Chemical Control

- Application of fungicides like fenhexamid and cyprodinil at key infection periods, particularly during bloom and pre-harvest.

Benefits of IPM

Environmental Benefits

1. **Reduced Chemical Usage:** IPM significantly reduces the reliance on chemical pesticides. By incorporating biological, cultural, and mechanical control methods, IPM minimizes the need for chemical interventions. This reduction in pesticide use lowers the risk of environmental contamination, including soil, water, and air pollution.
2. **Biodiversity Preservation:** The use of selective and reduced-risk pesticides, combined with biological control methods, helps preserve non-target species, including beneficial insects, birds, and soil microorganisms. This preservation of biodiversity is crucial for maintaining ecological balance and promoting natural pest control mechanisms.
3. **Enhanced Soil Health:** Cultural practices such as crop rotation, cover cropping, and organic amendments improve soil structure, fertility,

and microbial activity. Healthy soils support robust plant growth and resilience against pests and diseases, reducing the need for chemical inputs.

Economic Benefits

1. **Cost Efficiency:** IPM can lead to cost savings by reducing expenditures on chemical pesticides and mitigating crop losses due to pest damage. The investment in IPM practices, such as biological control agents and monitoring tools, often results in long-term economic benefits through improved crop yields and quality.

2. **Market Access and Premium Prices:** Crops produced using IPM practices often meet stringent pesticide residue standards required by domestic and international markets. This compliance can enhance market access and enable growers to achieve premium prices for their produce, particularly in markets that value sustainable and environmentally friendly practices.

3. **Sustainable Pest Control:** By promoting a diverse set of pest control methods, IPM reduces the risk of pest resistance to chemical pesticides. Sustainable pest control ensures long-term efficacy and stability in pest management, protecting the grower's investment and future yields.

Health Benefits

1. **Improved Food Safety:** IPM minimizes the use of chemical pesticides, leading to lower pesticide residues on crops. This improvement in food safety benefits consumers by reducing their exposure to potentially harmful chemicals in their diet.

2. **Worker Safety:** Reducing the frequency and volume of chemical pesticide applications decreases the risk of pesticide exposure for farm workers. IPM promotes safer working conditions, which can enhance labor productivity and reduce health-related absenteeism.

3. **Community Health:** Lower pesticide use reduces the risk of chemical drift and contamination of nearby communities, water sources, and ecosystems. This reduction protects public health and contributes to a safer living environment for local populations.

Challenges in IPM Adoption

1. **Knowledge and Awareness**: Limited knowledge and awareness of IPM principles and practices among farmers, extension workers,

and agricultural professionals pose a significant challenge. Many farmers may lack access to information on IPM techniques, including pest identification, monitoring methods, and alternative pest control strategies.

2. **Training and Capacity Building**: Effective implementation of IPM requires training and capacity building at various levels of the agricultural sector. Training programs on IPM may be lacking or insufficient, hindering farmers' ability to adopt and integrate IPM practices into their farming systems.

3. **Access to Resources**: Access to resources such as biopesticides, pheromone traps, and monitoring equipment can be limited, particularly for small-scale farmers in developing countries. The initial investment required to implement IPM practices, including the purchase of equipment and biological control agents, may be prohibitive for some farmers.

4. **Market and Policy Factors**: Market dynamics and policy frameworks can influence farmers' adoption of IPM. In some cases, market demands may prioritize quantity over quality, discouraging farmers from investing in IPM practices that enhance product quality but may increase production costs. Similarly, policy incentives and regulatory frameworks may not adequately support IPM adoption or provide incentives for sustainable pest management practices.

5. **Risk Aversion and Inertia**: Farmers may be hesitant to adopt new practices, particularly if they perceive IPM techniques as risky or unfamiliar. Traditional farming practices and reliance on chemical pesticides may be deeply ingrained, leading to inertia and resistance to change among farmers.

6. **Monitoring and Evaluation**: Effective IPM requires regular monitoring and evaluation of pest populations, environmental conditions, and the efficacy of control measures. However, monitoring can be labour-intensive and time-consuming, particularly for large-scale farming operations with limited manpower and resources.

7. **Pest Resistance**: The development of pest resistance to chemical pesticides remains a significant challenge in IPM implementation. Overreliance on specific pesticides can lead to the emergence of resistant pest populations, undermining the effectiveness of chemical control measures and necessitating alternative strategies.

8. **Coordination and Collaboration**: IPM implementation often requires coordination and collaboration among multiple stakeholders, including farmers, researchers, extension workers, and policymakers. Lack of coordination and communication among these stakeholders can hinder the adoption and scaling up of IPM practices.
9. **Scale and Context**: IPM practices may need to be adapted to suit local agroecological conditions, cropping systems, and pest pressures. Scaling up successful IPM interventions from pilot projects to larger agricultural landscapes can be challenging, requiring context-specific approaches and tailored solutions.
10. **Economic Viability**: While IPM offers long-term economic benefits through reduced pesticide use and improved crop yields, the initial investment and transition period may pose financial challenges for farmers, particularly those with limited resources or access to credit.

Conclusion

Integrated Pest Management (IPM) in horticultural crops is a sustainable and effective approach to pest control that balances ecological health, economic viability, and human safety. By integrating multiple control strategies and emphasizing monitoring and decision-making, IPM reduces the reliance on chemical pesticides and promotes long-term pest suppression. While challenges exist in knowledge dissemination, initial costs, and resistance management, the benefits of IPM—environmental preservation, economic efficiency, and improved health outcomes—make it a critical component of modern horticulture. Ongoing research, education, and policy support will be essential to further enhance the adoption and success of IPM practices globally.

References

Cameron, P. J. (2007). Factors influencing the development of integrated pest management (IPM) in selected vegetable crops: a review. New Zealand Journal of Crop and Horticultural Science, 35(3), 365-384.

Cuthbertson, A. G. (2020). Integrated pest management in arable and open field horticultural crops. Insects, 11(2), 82.

Nwilene, F. E., Nwanze, K. F., & Youdeowei, A. (2008). Impact of integrated pest management on food and horticultural crops in Africa. Entomologia experimentalis et applicata, 128(3), 355-363.

Mian, M. Y., Hossain, M. S., & Karim, A. R. (2016). Integrated pest management of vegetable crops in Bangladesh. Integrated pest management of tropical vegetable crops, 235-249.

Reddy, P. P. (2014). Biointensive integrated pest management in horticultural ecosystems. Berlin/Heidelberg, Germany: Springer.

Shankar, U. (2017). Integrated pest management in horticultural crops. Technological Innovations in Integrated Pest Management Biorational and Ecological Perspective, 307.

10

Integrated Disease Management in Agricultural Crops

Ektedar

Plant Pathology, Department of Plant Protection, Chaudhary Charan Singh University, Meerut, Uttar Pradesh

Abstract

Integrated Disease Management (IDM) in agricultural crops represents a comprehensive approach to mitigating the impact of plant diseases while ensuring sustainable crop production. The concept of IDM encompasses a range of strategies aimed at preventing, suppressing, or managing diseases through a combination of cultural, biological, chemical, and genetic control measures. This holistic approach acknowledges the complexity of disease dynamics within agroecosystems and emphasizes the integration of multiple tactics to minimize reliance on any single method. Cultural practices such as crop rotation, sanitation, and planting resistant cultivars form the foundation of IDM, reducing disease pressure and creating unfavourable conditions for pathogen development. Biological control agents, including beneficial microbes and natural enemies of plant pathogens, play a crucial role in regulating disease populations and promoting plant health. Chemical control measures, such as the judicious use of fungicides and bactericides, are employed when necessary but are integrated with other strategies to maximize efficacy and minimize environmental impact. Advances in molecular genetics have also contributed to IDM through the development of disease-resistant crop varieties with enhanced genetic resistance to specific pathogens. Overall, IDM offers a sustainable and multifaceted approach to disease management, addressing the challenges posed by plant pathogens while promoting environmental stewardship and long-term agricultural sustainability.

Keywords: *pathogens, cultivars, molecular, dynamics, biological*

Introduction

In the field of agriculture, integrated disease management (IDM) refers to a technique that is both comprehensive and sustainable. Its purpose is to successfully control plant diseases while simultaneously reducing negative consequences on the environment and assuring long-term output. This is a plan that is based on scientific research and incorporates a variety of illness management strategies in order to reach the best potential result. IDM seeks to achieve a balance between the economic viability of the project, the preservation of the environment, and the well-being of the community.

In order to keep plant populations healthy and produce maximum yields, integrated disease management (IDM) places an emphasis on prevention, early diagnosis, and the use of a number of methods. This strategy is especially important in contemporary agriculture, where reducing the quantity of chemical inputs and fostering techniques that are environmentally friendly are of the utmost importance.

Introduction to the History of Integrated Disease Management (IDM)

The history of Integrated Disease Management (IDM) may be traced back to the development of agricultural techniques and the comprehension of plant diseases. The formalization and development of integrated disease management (IDM) as a systematic method started to take shape sometime in the 20th century, despite the fact that the idea of integrating different techniques for illness management has been around for a very long time.

Early Agricultural measures: Throughout the course of human history, farmers have used a wide range of measures to control plant diseases. These activities often included aspects of what we now refer to as integrated disease management (IDM). A number of common strategies were used in order to lessen the effect of diseases, including crop rotation, intercropping, and the selection of disease-resistant species. The emergence of contemporary plant pathology as a scientific study may be traced back to the latter half of the 19th century and the opening decades of the 20th century. The causes, processes, and preventative measures of plant diseases were the subjects of researchers' first comprehensive investigations. The extensive use of synthetic pesticides for disease control gained pace in the middle of the 20th century, which led to a shift away from chemical control and toward integrated approaches. On the other hand, the difficulties that arise from depending only on chemical control, such as the development of resistance to pesticides and worries about the environment, became apparent.

In the 1960s and 1970s, the IDM was first developed.

- During this time period, researchers and practitioners began to recognize the limits of pharmacologic control and the need of adopting a more holistic approach.
- The notion that led to the development of the term "Integrated Pest Management" (IPM) was the idea of combining different approaches to the control of pests and diseases. These approaches included biological, cultural, and chemical approaches. The term "integrated control" was first introduced by entomologists from the University of California, which is where it got its start.

In the 1980s and 1990s, IDM was expanded.

Integrated Disease Management (IDM) started to be recognized as a separate subset of Integrated Pest Management (IPM), with a particular emphasis on the management of plant diseases.

- In an effort to successfully manage plant diseases, researchers and agricultural extension agencies started advocating for the integration of a variety of treatments, including resistant cultivars, cultural practices, and biological management.

The 21st Century: Developments and an Emphasis on Environmental Responsibility

- Since the beginning of the 21st century, there has been a growing focus on sustainable agriculture and the part that IDM plays in accomplishing this goal.
- As a result of widespread worries over the environment, the safety of food, and the effects of chemicals, there has been a renewed focus placed on integrated methods for the treatment of diseases.

Current Situation and Prospective Directions

- In today's contemporary agriculture, integrated disease management (IDM) has become a common technique. This approach emphasizes the necessity of integrating multiple tactics to successfully control plant diseases while reducing negative affects on both the environment and human health.

The creation of new tools and methods for integrated disease management is being aided by ongoing research and technology breakthroughs, which continue to contribute to the process.

Value of a Holistic Approach to Disease Control

- Reduced usage of chemical pesticides and increased use of ecologically friendly disease control approaches are two ways in which IDM supports sustainable agriculture. Agroecosystem health, biodiversity, and natural resource preservation are all improved by this.
- Hardiness in the Face of Disease Outbreaks: IDM builds a strong and hardy agricultural system by integrating several disease control strategies. The effect of some diseases on whole harvests may be lessened by cultural practices and crop diversification.
- Cultural methods, biological controls, and resistant plant types may all be part of IDM's strategy to drastically cut down on chemical pesticide use. Soil, water, crops, and ecosystems are all spared from chemical residues, which is good news for people and the planet.
- Maintaining or enhancing crop yields and quality while decreasing the total cost of managing illnesses is the goal of integrated disease management (IDM), which strives to optimize disease management tactics and improve economic viability. As a result, farming becomes more financially feasible.
- Benefit to the Environment: IDM helps lessen pollution and protects important creatures like pollinators and natural pest controllers by lowering the usage of chemical pesticides.
- Health and Safety for All: IDM places an emphasis on using pesticides in a responsible and safe manner, which helps to protect agricultural workers, consumers, and the environment from harm.
- Scientific Advancement: New technology, biological control agents, and resistant crop varieties are all products of the research and innovation spurred by the deployment of integrated disease management (IDM).
- In order to respond to the difficulties presented by climate change and new diseases, it is vital to have a diverse and adaptive agricultural system. IDM supports just that.
- The biological control strategies used in integrated pest management (IDM) help to protect natural enemies of pests and diseases, which in turn helps to maintain a healthy environment. A more robust agroecosystem and less secondary pest outbreaks are possible outcomes of this.
- Integrated Approach: IDM takes a comprehensive approach that acknowledges the intricate nature of disease relationships. Integrated

disease management (IDM) integrates many instruments to customize disease management techniques for different crops, geographies, and conditions, instead of depending on a single control strategy.

A Plant Disease Concept

Tree, grain, and legume illnesses were originally documented by the Greek philosopher Theophrastus (300 B.C.). He thought that the weather, which carried illnesses, was under God's authority. The value of God was shown by plant diseases. This may be the result of superstitions, religious beliefs, or even the influence of the moon and harsh winds. To combat rust illnesses in grain harvests, the Romans, for instance, established a deity specifically for the purpose, Robigo. Red dogs and lambs were given as sacrifices. Mildews, rust, and other symptoms seen on plants and microorganisms discovered on sick plants were first linked with the introduction of the compound microscope in the mid-seventeenth century, which allowed scientists to view several bacteria linked with sick plants. There is incontestable proof that microbes can only evolve from other microbes and that fermentation is a biological, not just chemical, phenomena, according to the work of Louis Pasteur (1860–63). When a plant is able to do all of its physiological tasks to the best of its genetic ability, we say that it is healthy or normal. Pathogens may inflict plant diseases in a few different ways: -Depleting the host plant by sucking its nutrients out of its cells all the time. -Secreting poisons, enzymes, or growth-regulating compounds that kill or disrupt the metabolism of host cells. -Eating the contents of the host cells upon contact. -Blocking the transfer of food, minerals, nutrients, and water via the conductive tissues.

Triangle of Disease

The "disease triangle" is a common way to represent the interplay between the host, the pathogen, and the physical environment (including climate, soil, geography, and biology). Figure 1 shows the three parts of a triangle, with each side representing one of them. Each side's length is directly proportionate to the total of the components' disease-promoting traits. For example, if the host plant is resistant, fully grown, and widely spread, the host side would be short and the disease quantity would be little or nonexistent. On the other hand, if the host plant is sensitive, in the vulnerable development stage, or densely planted, the host side would be long and the disease amount may be large. The length of time an illness persists and the severity of its symptoms are both affected by the pathogen's virulence, abundance, and activity. In addition, the environmental factors that contribute to the pathogen's success or that weaken the host's resistance—such as temperature, moisture, and wind—determine the duration of the environmental side and the severity of the illness. A plant's or population's disease load is proportional to the area of the triangle formed by these three disease indicators. There can be no illness if any one of the three parts is zero.

Instruments Used in Integrated Disease Management

The instruments used in Integrated Disease Management (IDM) include a diverse range of strategies that are employed to efficiently manage and regulate plant diseases, all the while fostering sustainable agricultural practices. These tools consist of many tactics and approaches to tackle difficulties associated to diseases. Below are the essential instruments used in IDM:

1. Resistant Host Plant Varieties

- Utilizing plant types that possess inherent resistance to certain diseases is a core strategy in Integrated Disease Management (IDM). The process of breeding for resistance is the deliberate selection and cultivation of plant varieties that possess genetic characteristics that provide immunity or defense against harmful microorganisms.
- Utilizing resistant cultivars may serve as a straightforward, practical efficient, and cost-effective approach to managing plant diseases.
- In addition to providing disease protection, they may also save time, money, and energy that would otherwise be spent on other control measures, while also preventing chemical contamination of the environment.
- Biological control methods are the only effective means of managing diseases like as wilts, rusts, and viral infections, when chemical control is prohibitively costly and unfeasible.
- For low-value crops, when alternative approaches are sometimes too costly, it may be advisable for farmers to cultivate types that are resistant to common and significant illnesses.
- Plant disease resistance is determined by their genetic makeup and may be categorized as monogenic, oligogenic, or polygenic. Origin

2. Cultural Practices: Cultural practices have a crucial role in preventing diseases. These strategies include crop rotation, appropriate spacing, maximizing planting density, modifying irrigation procedures, and establishing effective sanitation measures.

- **Intensive tillage:** Deep plowing of the field leads to the propagules being exposed to higher temperatures and physically eliminates the pathogen. This phenomenon may be considered as dry soil solarization.
- The act of plowing throughout the summer season was shown to be successful in decreasing the number of cyst nematodes and enhancing the productivity of wheat crops.
- **Field inundation:** The flooding of the field has some resemblance to soil disinfestation. It has been shown that long-term summer soil flooding, whether or not accompanied by paddy cultivation, leads to a drop in populations of soil borne diseases.
- Crop rotation and diversification include the practice of alternating the cultivation of different kinds of crops in a field. This strategy may effectively interrupt the life cycle of pathogens, hence minimizing the occurrence and impact of diseases. Diversification enhances the equilibrium of an ecosystem and mitigates the accumulation of certain diseases.
- Residue management involves implementing practices such as the removal and proper disposal of contaminated plant waste, the incorporation of agricultural residues into the soil by plowing, and the use of crop rotation. These practices are effective in reducing the survival and transmission of infections.

Additional Effective Cultural Practices

- To minimize the spread of soil-borne viruses across fields, it is necessary to disinfect stakes and agricultural equipment prior to transporting them from one field to another.
- Minimize soil displacement between sites to decrease the likelihood of disease transfer.
- Effective weed control is crucial for the management of viral diseases. • Certain infections can only infect the host via wounds, thus, circumstances that lead to plant harm should be prevented.
- The amount of disease-causing organisms may be decreased by eliminating both sick and healthy plant material after the harvest.

3. Biological Control Agents

- Predators, parasitoids, and beneficial microorganisms are used as helpful creatures to manage disease-causing species. These indigenous predators assist in regulating the populations of diseases, hence fostering a harmonious ecology.

- Biocontrol agents serve as a fundamental element of an integrated disease management strategy. (For instance: Bacillus subtilis, Pseudomonas fluorescens, Gliocladium spp., and other similar examples)
- Biopesticides, which are produced from natural sources such as fungus, bacteria, and plant extracts, are used to control pests. They provide a substitute for synthetic chemical pesticides and may be efficient in managing illnesses.
- Predators and parasites play a crucial role in maintaining the balance of insect populations, since they prey on and control the numbers of harmful insects. Similarly, some insects and mites have the ability to consume and control plant infections. Predatory nematodes engage in attacking detrimental nematodes present in the soil, while parasitic wasps deposit their eggs into pest insects. Competition may be induced by introducing non-pathogenic strains of a disease-causing bacteria, which leads to a scarcity of resources and thus makes it more difficult for the harmful strain to establish itself.
- Induced Resistance: Beneficial bacteria have the ability to activate the plant's innate defense systems, enhancing its resistance to illnesses.

4. Physical Control

This entails using physical barriers or techniques to hinder the transmission of illnesses. Some examples of methods to restrict the spread of diseases include using row covers, implementing mulching techniques, and practicing sanitary measures.

- Gather and eliminate the plant parts that are afflicted with the illness.
- Soil solarization is a method used to manage soil borne illnesses caused by fungi that are often difficult to treat, such as Rhizoctonia solani, Fusarium spp., and Sclerotium.
- The soil beds are watered first and then coated with a thin (20 μm) translucent mulch throughout the months of April, May, and June. In many instances, it elevated the soil temperatures by as much as 50°C, which is detrimental to several soil-borne plant diseases. It has been used in establishing disease-free nurseries in tropical and subtropical regions. Additionally, it offers exceptional weed management.

- **Hot water treatment:** Seed borne infections may be effectively treated using hot water therapy, which involves immersing contaminated seeds in water at the correct temperature and duration. Immersing cabbage seed in hot water at a temperature of 52°C for a duration of 15-20 minutes effectively manages black rot disease, which is caused by Xanthomonas campestris pv. campestris.
- **Hot air therapy:** It is used to eliminate excessive moisture from plant organs and safeguard them from fungal and bacterial infections. Dormant plants infected with several viruses are subjected to hot air treatment at temperatures ranging from 35-54°C for a duration of 8 hours.
- Refrigeration, often known as low temperature treatment, is the most widely used way to prevent postharvest infections in perishable fruits and vegetables.
- **Solar thermal therapy:** Exposing water-soaked wheat seeds to solar heat treatment for 5-6 hours in May-June effectively controls loose smut disease in wheat.
- Many post-harvest diseases may be prevented by methods such as irradiation, refrigeration, and Controlled Atmosphere Storage.

5. Chemical Control (As a Last Resort): If other approaches are not enough, certain chemical pesticides might be utilized. To get optimal results, it is crucial to use chemicals in a careful and thoughtful manner, taking into account elements such as the particular target, the time of administration, and the possible effects on non-target species.

Chemical control encompasses the use of many categories of chemicals

- **Fungicides:** Employed to specifically combat fungal illnesses.
- **Bactericides:** Specifically designed to combat bacterial infections.
- Insecticides specifically target and combat insect infestations.
- **Nematicides:** Designed to specifically target and control nematode pests.
- **Virucides:** Designed to specifically combat viral illnesses (restricted selection).

Seed treatment is the application of fungicides, biological agents, or other treatments to seeds prior to planting. This practice offers early protection against diseases that are present in the soil.

Disease thresholds are the specific levels at which it becomes important to take action in order to avert economic harm caused by the disease. These standards assist farmers in making well-informed choices on disease management strategies.

6. Quarantine and regulatory measures are crucial components of Integrated Disease Management (IDM) since they aim to prevent the entry and dissemination of foreign infections by implementing rigorous restrictions and quarantine protocols.

- Plant quarantine refers to the legally mandated limitation on the transportation of plant materials that are infected or carry fungus, bacteria, or viruses that might cause plant diseases.
- Quarantine and regulatory measures are implemented as a means of excluding and controlling plant diseases, in accordance with the Principles of Plant Disease Control.
- Through the use of these techniques, IDM strives to develop a holistic, efficient, and enduring strategy for the management and regulation of plant diseases, with the goal of reducing adverse effects on the environment and human well-being.

Conclusion

In conclusion, Integrated Disease Management (IDM) represents a proactive and sustainable approach to addressing the complex challenges posed by plant diseases in agricultural crops. By integrating diverse strategies including cultural, biological, chemical, and genetic controls, IDM aims to minimize disease incidence and severity while preserving environmental quality and ensuring long-term crop productivity. While individual components of IDM may vary depending on crop, region, and specific disease pressures, the overarching principle remains consistent: to employ a holistic and integrated approach that maximizes effectiveness while minimizing reliance on any single method. As agriculture continues to face evolving disease threats and environmental concerns, IDM offers a path forward towards resilient and sustainable crop production systems. Continued research, education, and collaboration among stakeholders will be essential in further advancing IDM practices and ensuring their widespread adoption for the benefit of farmers, consumers, and the environment.

References

Abbas, M., Saleem, M., Hussain, D., Ramzan, M., Jawad Saleem, M., Abbas, S., ... & Parveen, Z. (2022). Review on integrated disease and pest management of field crops. International Journal of Tropical Insect Science, 42(5), 3235-3243.

Kakraliya, S. S., Abrol, S., Choskit, D., & Pandit, D. (2020). Integrated Disease Management in Agriculture. Just Agriculture, 1(3), 17.

Khan, S. M., Ali, S., Nawaz, A., Bukhari, S. A. H., Ejaz, S., & Ahmad, S. (2019). Integrated pest and disease management for better agronomic crop production. Agronomic Crops: Volume 2: Management Practices, 385-428.

Khoury, W. E., & Makkouk, K. (2010). Integrated plant disease management in developing countries. Journal of Plant Pathology, S35-S42.

Pandey, A. K., Sain, S. K., & Singh, P. (2016). A Perspective on integrated disease management in agriculture. Bio Bulletin, 2(2), 13-29.

Razdan, V. K., & Sabitha, M. (2009). Integrated disease management: Concepts and practices. Integrated Pest Management: Innovation-Development Process: Volume 1, 369-389.

11

Integrated Disease Management in Horticultural Crops

***Riya Thakur*[1] *and Ajay Haldar*[2]**

[1]*Scientist (Horticulture), Krishi Vigyan Kendra, Chhindwara, M.P.*
[2]*Assistant Professor (Horticulture), G.H. Raisoni University, Saikheda, M.P.*

Abstract

Integrated Disease Management (IDM) in horticultural crops is a multifaceted approach aimed at mitigating the impact of plant diseases while ensuring sustainable production systems. Recognizing the complex interactions between pathogens, plants, and the environment, IDM integrates various control measures, including cultural, biological, chemical, and genetic strategies. Cultural practices such as crop rotation, proper sanitation, and selection of disease-resistant cultivars form the foundation of IDM, reducing disease pressure and creating unfavourable conditions for pathogen development. Biological control agents, including beneficial microbes and natural enemies of plant pathogens, play a crucial role in regulating disease populations and promoting plant health. Chemical interventions, when necessary, are judiciously applied and integrated with other strategies to minimize environmental impact and preserve beneficial organisms. Advances in molecular genetics have also contributed to IDM through the development of disease-resistant crop varieties with enhanced genetic resistance to specific pathogens. Overall, IDM offers a holistic and sustainable approach to disease management in horticultural crops, emphasizing proactive measures to prevent disease outbreaks and minimize the need for reactive interventions. Continued research, education, and collaboration among stakeholders are essential for advancing IDM practices and ensuring their effective implementation in horticultural production systems.

Keywords: *horticultural, judiciously, population, plant, advances*

Introduction to Integrated Disease Management (IDM)

IDM involves the use of diverse strategies to manage plant diseases in a coordinated manner. The primary goal is to achieve effective disease control while minimizing the environmental impact and enhancing crop productivity and quality. IDM strategies include:

Crop Rotation

Crop rotation involves growing different types of crops in the same area in sequential seasons. This practice helps break the life cycles of pathogens and pests that specialize in specific crops, reducing their population in the soil. For example, rotating crops from different plant families can effectively manage soil-borne diseases like Fusarium wilt and nematodes. Including non-host crops in the rotation disrupts the continuous availability of the host plant required for the pathogen's survival.

Sanitation

Sanitation practices involve the removal and destruction of plant debris, weeds, and other sources of pathogen inoculum. By eliminating infected plant material, the spread of diseases can be minimized. This includes:

- **Pruning infected plant parts**: Regularly removing diseased leaves, stems, and fruits to prevent the spread of pathogens.
- **Disinfecting tools and equipment**: Cleaning and disinfecting pruning tools, harvesting equipment, and other implements to avoid cross-contamination.

Irrigation Management

Proper irrigation techniques are crucial to prevent diseases, especially those caused by waterborne pathogens. Some effective irrigation management practices include:

- **Drip Irrigation**: Applying water directly to the soil around the plants' root zone, reducing leaf wetness and the likelihood of foliar diseases.
- **Avoiding Overhead Irrigation**: Minimizing the use of overhead sprinklers that can increase humidity and create conditions favourable for fungal and bacterial infections.
- **Timely Watering**: Watering plants early in the day to allow foliage to dry before nightfall, reducing the duration of leaf wetness and the risk of disease development.

Soil Management

Soil health is fundamental to crop health. Practices that improve soil structure and fertility can enhance plant resistance to diseases. These include:

- **Organic Amendments**: Adding compost, manure, and other organic materials to improve soil structure, increase microbial activity, and enhance nutrient availability.
- **Cover Cropping**: Growing cover crops such as legumes and grasses to improve soil fertility, suppress weeds, and reduce erosion. Cover crops can also promote beneficial soil microorganisms that compete with or inhibit pathogens.

Pruning and Training

Pruning and training plants to improve air circulation and light penetration within the canopy can significantly reduce the incidence of diseases. This involves:

- **Removing Excess Foliage**: Thinning out dense foliage to allow better air movement and light penetration, reducing humidity levels and the potential for fungal infections.
- **Supporting Plants**: Using stakes, trellises, or cages to support plants and keep fruits and leaves off the ground, minimizing contact with soil-borne pathogens.

Resistant Varieties

Selecting and planting disease-resistant or tolerant varieties is one of the most effective cultural practices in IDM. Breeding programs have developed varieties with genetic resistance to specific diseases, reducing the need for chemical interventions. Farmers are encouraged to choose varieties that are well-adapted to local conditions and resistant to prevalent diseases in the area.

Mulching

Mulching involves covering the soil with organic or inorganic materials to conserve moisture, suppress weeds, and reduce soil-borne diseases. Benefits of mulching include:

- **Moisture Retention**: Keeping the soil moist and reducing water stress on plants.
- **Temperature Regulation**: Moderating soil temperatures, which can be beneficial in both hot and cold climates.
- **Weed Suppression**: Reducing competition from weeds that can harbour pests and diseases.

- **Barrier Against Soil Pathogens**: Creating a physical barrier that prevents soil-borne pathogens from splashing onto plants during rain or irrigation

Cultural Practices

Cultural practices are essential components of Integrated Disease Management (IDM) in horticultural crops. These practices aim to reduce the incidence and severity of diseases by creating unfavourable conditions for pathogens and enhancing the natural resilience of plants. Here is a detailed overview of key cultural practices used in IDM:

Crop Rotation

Crop rotation involves growing different crops in the same area across different seasons. This practice disrupts the life cycles of pests and pathogens that specialize in particular crops. For example, alternating between root crops and leafy vegetables can help manage soil-borne diseases like Fusarium wilt and nematodes by depriving pathogens of their preferred hosts.

Sanitation

Sanitation is crucial for preventing the spread of diseases in horticultural crops. Effective sanitation practices include:

- **Removal of Diseased Plants**: Uprooting and destroying infected plants and plant debris to reduce pathogen reservoirs.
- **Cleaning Tools and Equipment**: Regularly disinfecting tools, containers, and equipment to prevent the spread of pathogens between plants and fields.
- **Managing Crop Residues**: Properly disposing of crop residues or incorporating them into the soil after composting to reduce disease inoculum.

Irrigation Management

Proper irrigation techniques can significantly reduce the risk of disease by managing the moisture levels that pathogens require to thrive. Effective irrigation management practices include:

- **Drip Irrigation**: This method delivers water directly to the plant roots, reducing leaf wetness and the spread of waterborne pathogens.
- **Avoiding Overhead Irrigation**: Overhead sprinklers can increase humidity and leaf wetness, promoting fungal and bacterial diseases. Drip or furrow irrigation is preferred.

- **Water Timing**: Watering early in the day allows foliage to dry quickly, reducing the duration of leaf wetness and the risk of disease development.

Soil Management

Healthy soil is fundamental for disease prevention and plant health. Practices that enhance soil health include:

- **Organic Amendments**: Adding compost, manure, and other organic materials improves soil structure, increases microbial activity, and enhances nutrient availability.
- **Cover Crops**: Growing cover crops like legumes and grasses can improve soil fertility, suppress weeds, and reduce soil erosion. They also support beneficial soil microorganisms that compete with pathogens.
- **Soil Testing and Amendments**: Regular soil testing to monitor nutrient levels and pH can help tailor fertilization and amendment practices to maintain optimal soil conditions for crop growth.

Pruning and Training

Pruning and training plants improve air circulation and light penetration within the crop canopy, reducing humidity and the potential for disease. Key practices include:

- **Removing Excess Foliage**: Thinning dense plant growth to allow better air movement and light penetration, which helps in drying out plant surfaces and reducing fungal infections.
- **Proper Plant Spacing**: Ensuring adequate spacing between plants to minimize overcrowding, which can create a microclimate conducive to disease development.
- **Supporting Plants**: Using stakes, trellises, or cages to keep plants and fruits off the ground, minimizing contact with soil-borne pathogens.

Resistant Varieties

Planting disease-resistant or tolerant varieties is a proactive way to manage diseases. Breeding programs focus on developing cultivars with resistance to specific pathogens, reducing the need for chemical controls. Selecting varieties well-suited to local conditions and known to resist prevalent diseases can significantly reduce disease pressure.

Mulching

Mulching involves covering the soil with organic or inorganic materials to suppress weeds, retain soil moisture, and reduce soil-borne diseases. Benefits of mulching include:

- **Moisture Conservation**: Mulch helps retain soil moisture, reducing water stress on plants.
- **Temperature Regulation**: Mulch moderates soil temperatures, protecting roots from extreme heat or cold.
- **Weed Suppression**: Mulch reduces weed growth, which can harbour pests and diseases.
- **Barrier Against Soil Pathogens**: Mulch prevents soil-borne pathogens from splashing onto plants during rain or irrigation, reducing disease transmission.

Biological Control

Biological control is a key component of Integrated Disease Management (IDM) in horticultural crops. It involves using living organisms to control plant pathogens and pests. This approach aims to reduce the reliance on chemical pesticides, promoting a more sustainable and environmentally friendly agricultural system. Here, we explore various aspects of biological control, including the use of beneficial microorganisms, predators, parasitoids, and biopesticides.

Beneficial Microorganisms

Beneficial microorganisms, including bacteria and fungi, play a crucial role in biological control by antagonizing plant pathogens through competition, antibiosis, or parasitism. Some well-known examples include:

- **Trichoderma spp.**: These fungi are effective against several soil-borne pathogens. They colonize the root zone, outcompeting pathogens for space and nutrients, and produce enzymes and antibiotics that inhibit pathogen growth. Trichoderma also promotes plant growth by enhancing nutrient uptake.
- **Bacillus spp.**: Bacillus species, such as Bacillus subtilis, produce antibiotics and enzymes that degrade the cell walls of pathogenic fungi and bacteria. They also induce systemic resistance in plants, enhancing their defence mechanisms against a broad range of pathogens.

Predators and Parasitoids

Natural predators and parasitoids are essential in managing insect pests that act as vectors for plant diseases. Examples include:

- **Lady Beetles (Coccinellidae)**: These beetles are voracious predators of aphids, mites, and other small insects that can transmit plant diseases. Their presence in the garden helps reduce pest populations naturally.
- **Parasitic Wasps (e.g., Trichogramma spp.)**: These wasps lay their eggs inside the eggs or larvae of pests like caterpillars and flies. The developing wasp larvae consume the host, effectively reducing pest numbers. This method is particularly useful in controlling lepidopteran pests in crops.

Biopesticides

Biopesticides are derived from natural organisms or their by-products. They offer a targeted approach to pest and disease management with minimal impact on non-target species and the environment. Examples include:

- **Bt (Bacillus thuringiensis)**: This bacterium produces toxins that are specifically lethal to certain insect larvae, such as caterpillars, while being harmless to humans, animals, and beneficial insects. Bt is widely used in organic farming as a spray or soil treatment.
- **Neem Oil**: Extracted from the neem tree, neem oil contains compounds that disrupt the life cycle of insects, acting as an insect repellent, growth regulator, and feeding deterrent. It is effective against a variety of pests and has antifungal properties as well.

Benefits of Biological Control

- **Sustainability**: Biological control reduces the need for chemical pesticides, lowering the risk of environmental pollution and promoting biodiversity.
- **Safety**: Most biocontrol agents are specific to their target pests or pathogens, posing little or no risk to humans, animals, and beneficial insects.
- **Resistance Management**: Biological control agents typically work through multiple mechanisms, making it less likely for pests and pathogens to develop resistance compared to chemical pesticides

Chemical Control

Physical control methods are vital components of Integrated Disease Management (IDM) in horticultural crops, focusing on the use of physical barriers, treatments, and techniques to manage plant diseases and pests. These methods are environmentally friendly, reduce the reliance on chemical pesticides, and can be integrated with other IDM strategies for comprehensive disease management. Here's an overview of various physical control techniques used in horticultural crops:

Soil Solarization

Soil solarization involves covering the soil with transparent polyethylene sheets to trap solar radiation, thereby heating the soil to levels that are lethal to many soil-borne pathogens, nematodes, and weed seeds. The process typically takes four to six weeks during the hottest part of the year. Benefits include:

- **Pathogen Reduction**: Effectively reduces populations of fungi, bacteria, nematodes, and weeds.
- **Enhanced Soil Health**: Can improve soil structure and nutrient availability by killing harmful soil organisms while promoting beneficial ones.

Mulching

Mulching involves covering the soil with organic or inorganic materials to suppress weeds, retain soil moisture, and reduce soil-borne diseases. Types of mulches and their benefits include:

- **Organic Mulches**: Straw, wood chips, and compost that decompose over time, adding organic matter to the soil and improving soil health.
- **Inorganic Mulches**: Plastic films and landscape fabrics that provide immediate physical barriers to weed growth and soil pathogen contact.
- **Weed Suppression**: Mulch reduces weed emergence, which can otherwise harbour pests and diseases.
- **Moisture Conservation**: Helps maintain consistent soil moisture, reducing water stress on plants.

Thermal Treatments

Thermal treatments involve the use of heat to kill pathogens and pests. Methods include:

- **Hot Water Treatment**: Soaking seeds, tubers, or cuttings in hot water to eliminate seed-borne pathogens without affecting germination.
- **Steam Sterilization**: Using steam to sterilize soil, substrates, or tools to kill pathogens and pests. This is particularly useful in greenhouse settings where soil-borne diseases can be a significant problem.

Barriers and Traps

Physical barriers and traps can effectively prevent pest access to crops and reduce disease transmission. Examples include:

- **Row Covers**: Lightweight fabric covers that protect plants from insect pests while allowing light, air, and moisture to penetrate. These covers can prevent pests like aphids, beetles, and caterpillars from reaching the plants.
- **Insect Traps**: Sticky traps, pheromone traps, and light traps can be used to monitor and reduce pest populations. These traps help in early detection and control of pest outbreaks.
- **Physical Barriers**: Using fences, nets, or screens to exclude larger pests like birds, rabbits, and deer from accessing the crops.

Pruning and Canopy Management

Proper pruning and canopy management improve air circulation and light penetration, reducing the humidity levels that favour many fungal and bacterial diseases. Techniques include:

- **Selective Pruning**: Removing diseased or overcrowded branches to improve airflow and reduce disease pressure.
- **Training Systems**: Implementing trellising and staking systems to keep plants off the ground and reduce their contact with soil-borne pathogens .

Water Management

Effective water management practices can significantly reduce the risk of disease by controlling moisture levels. These practices include:

- **Drip Irrigation**: Delivers water directly to the plant roots, reducing leaf wetness and the spread of foliar diseases.
- **Avoiding Overhead Irrigation**: Minimizes the risk of spreading waterborne pathogens and reduces humidity around plants.

- **Proper Drainage**: Ensuring fields and containers have adequate drainage to prevent waterlogging, which can promote root diseases.

Soil Amendments and Conditioning

Applying soil amendments and conditioners can enhance soil health and reduce disease incidence. Examples include:

- **Compost and Organic Matter**: Adding organic materials to improve soil structure, water retention, and microbial diversity, which can suppress soil-borne pathogens.
- **pH Adjustment**: Adjusting soil pH to levels that are unfavourable for specific pathogens but optimal for crop growth.

Case Studies and Examples

Integrated Disease Management (IDM) in horticultural crops involves a combination of cultural, biological, physical, and chemical control methods to effectively manage diseases. Here are some case studies and examples illustrating the successful implementation of IDM strategies in various horticultural crops:

1. IDM in Tomato Production

Background

Tomatoes are susceptible to a wide range of diseases, including bacterial wilt, early blight, and Fusarium wilt. Managing these diseases requires an integrated approach due to their complex nature and varied modes of transmission.

Strategies and Outcomes

- **Cultural Practices**: Crop rotation with non-host crops (such as cereals) significantly reduced soil-borne pathogens. Raised bed cultivation improved soil drainage, reducing the incidence of root rot diseases.
- **Biological Control**: Application of **Trichoderma harzianum** and **Bacillus subtilis** as soil amendments helped suppress Fusarium wilt by outcompeting the pathogen and enhancing plant resistance.
- **Physical Control**: The use of plastic mulches suppressed weeds and prevented soil pathogen splash during irrigation. Regular pruning and staking of plants improved air circulation, reducing the humidity that favours fungal diseases.
- **Chemical Control**: Fungicides such as chlorothalonil were used in a targeted manner, based on disease monitoring and threshold levels.

Copper-based bactericides were applied to manage bacterial wilt during early stages of infection.

Results

The integrated approach led to a significant reduction in disease incidence and improved overall plant health and yield. Farmers observed a 40% decrease in fungicide use, demonstrating the effectiveness and sustainability of IDM strategies.

2. IDM in Apple Orchards

Background

Apple orchards are affected by several diseases, including apple scab, fire blight, and powdery mildew. Effective IDM is crucial to maintain high fruit quality and yield.

Strategies and Outcomes

- **Cultural Practices**: Pruning and training systems were employed to improve air circulation and reduce disease pressure. Fallen leaves and debris, which can harbour apple scab spores, were regularly removed and composted.
- **Biological Control**: The introduction of **Bacillus pumilus** as a foliar spray helped control apple scab and powdery mildew. Natural predators, such as lady beetles, were encouraged to control aphid populations, reducing the spread of viral diseases.
- **Physical Control**: Reflective ground covers were used to reduce aphid populations by disorienting them, thereby lowering the incidence of aphid-transmitted viruses. Netting was applied to protect against insect vectors of fire blight.
- **Chemical Control**: Systemic fungicides like myclobutanil were applied during critical periods, such as petal fall, to prevent apple scab. Antibiotics like streptomycin were used judiciously to manage fire blight, following strict guidelines to prevent resistance.

Results

The implementation of IDM strategies led to a marked decrease in apple scab and fire blight incidences. The integrated approach allowed for a 30% reduction in chemical pesticide use, enhancing the sustainability of apple production.

3. IDM in Strawberry Fields

Background

Strawberries are prone to diseases such as Botrytis fruit rot (gray mold), powdery mildew, and Verticillium wilt. An integrated approach is essential to manage these diseases effectively.

Strategies and Outcomes

- **Cultural Practices**: Crop rotation with non-host plants like legumes and grasses helped break the disease cycle. Raised beds and proper spacing reduced soil moisture and improved air circulation.
- **Biological Control**: **Bacillus subtilis** and **Trichoderma spp.** were applied as soil amendments and foliar sprays to manage Botrytis and soil-borne diseases. Predatory mites were released to control spider mite populations, which can exacerbate disease problems.
- **Physical Control**: Straw mulch was used to reduce soil splash and maintain fruit cleanliness, reducing the incidence of gray mold. Row covers protected plants from frost and insect vectors.
- **Chemical Control**: Fungicides such as fenhexamid were applied during the flowering and fruiting stages to prevent Botrytis infections. The use of these fungicides was based on disease forecasting models and regular monitoring.

Results

The combination of cultural, biological, physical, and chemical controls resulted in a substantial reduction in disease pressure and improved fruit quality.

Conclusion

Integrated Disease Management in horticultural crops represents a holistic approach to plant health that prioritizes sustainability and environmental stewardship. By integrating cultural, biological, physical, and chemical methods, IDM reduces the risk of disease outbreaks and minimizes the negative impacts of conventional pest control practices. As the horticultural industry continues to grow and evolve, the adoption of IDM principles will be crucial for ensuring the long-term health and productivity of crops.

References

Ahamad, S., Anwar, A., & Sharma, P. K. (2011). Plant Disease Management on Horticultural Crops.

Akem, C. N., & Jovicich, E. (2013). Integrated management of foliar diseases in vegetable crops.

Das, D., & Baruah, M. (2010). Sustainable practices for pest and disease management of horticultural crops. Annals of Plant Protection Sciences, 18(2), 357-361.

Gajanana, T. M., Krishna Moorthy, P. N., Anupama, H. L., Raghunatha, R., & Kumar, G. T. (2006). Integrated pest and disease management in tomato: an economic analysis. Agricultural economics research review, 19(2), 269-280.

Garibaldi, A., & Gullino, M. L. (2009, September). Emerging soilborne diseases of horticultural crops and new trends in their management. In VII International Symposium on Chemical and Non-Chemical Soil and Substrate Disinfestation 883 (pp. 37-47).

Kurmi, S., Patidar, R., Kurmi, P., Patel, R., & Singh, Y. Integrated Approaches in Disease Management in Horticultural Crops. Revolutionizing Horticulture The Green Path, 137.

Narayanasamy, P., & Narayanasamy, P. (2013). Biological Disease Management Systems for Horticultural Crops. Biological Management of Diseases of Crops: Volume 2: Integration of Biological Control Strategies with Crop Disease Management Systems, 237-346.

12

Integrated Weed Management in Agricultural Crops

***Sanjeev Kumar*[1]*, Deovrat Singh*[2]*, Anushi*[3] *and Durgesh Kumar Maurya*[4]**

[1]*Department of Agronomy, Banaras Hindu University Varanasi, Uttar Pradesh*
[2]*Department of Agriculture, IIAST, Integral University, Lucknow Uttar Pradesh*
[3]*Department of Fruit Science, Chandra Shekhar Azad University of Agriculture & Technology, Kanpur, Uttar Pradesh*
[4]*Department of Agronomy, Chandra Shekhar Azad University of Agriculture & Technology, Kanpur, Uttar Pradesh*

Abstract

Integrated weed management (IWM) in agricultural crops is a comprehensive approach combining multiple strategies to control weed populations effectively and sustainably. IWM integrates cultural, mechanical, biological, and chemical methods to manage weeds in a way that minimizes the reliance on herbicides, reduces the risk of herbicide resistance, and promotes environmental health. Cultural practices include crop rotation, cover cropping, and optimizing planting densities to outcompete weeds. Mechanical methods involve tillage, mowing, and manual removal, while biological control utilizes natural predators and competitive plants. Chemical controls are applied judiciously, targeting specific weed issues with appropriate herbicides. By employing a diversified approach, IWM enhances crop productivity, supports biodiversity, and fosters long-term agricultural sustainability.

Keywords: *agricultural, optimizing, chemicals, controls*

Introduction

Integrated Weed Management (IWM) refers to the process of combining several strategies for controlling weeds into a single program, with the goal of maximizing the effectiveness of weed control for a specific weed issue. In recent decades, there has been a shift towards simpler weed management methods that mainly depend on a few number of widely-used herbicides. Nevertheless, the fast proliferation of herbicide-resistant weeds has compelled farmers to integrate alternate methods of weed control. Although some farmers are integrating various herbicides, the efficacy of this approach is expected to be temporary. For long-term effectiveness, it is necessary to use non-herbicide methods in conjunction with many, effective sites of action. Prioritizing a discussion on the need of weed management would be more advantageous. Weeds have a detrimental effect on agricultural yields, disrupt various crop production processes, and weed seeds may contaminate grain. According to nationwide study, the productivity of maize and soybean crops might be decreased by over 50% if appropriate measures to manage weeds are not implemented. The primary method used for weed management is the application of herbicides. The exclusive reliance on this particular strategy has resulted in the emergence of herbicide-resistant weeds. There is a limited selection of herbicides that may be used, and instances of herbicide resistance are increasing swiftly in the United States. Therefore, herbicides need additional assistance to maintain sufficient weed control. The strategies used by IWM include a broad spectrum of choices and intricacy. Several IWM strategies may be incorporated into existing management programs with few adjustments, while others need thorough design and execution. Some simpler solutions to apply include: cleaning equipment, scouting at the appropriate time, and adjusting herbicide tank mixtures. On the other hand, more comprehensive options include changing crop rotation, using cover cropping, modifying tillage methods, and implementing weed seed management around harvest time.

Preventive Measures

Taking proactive measures is crucial in effectively managing weeds. This category stands apart from the rest as it prioritizes the prevention of weeds from infiltrating or spreading within a field. Growers can implement this strategy by ensuring that your inputs are free from weed seeds, including crop seed, manure, and other materials. Equipment used for cleaning, such as combines, has the potential to inadvertently transport weed seeds from one field to another. Ensuring that weeds are unable to reproduce in the field as well as in ditches, fencerows, and other adjacent non-crop areas. Ensuring prompt identification and removal of any unwanted plants. It is important to exercise caution when buying second-hand farm equipment or leasing land for agricultural purposes.

Cultural Measures

The greatest method of weed management is a crop that is robust and healthy. The purpose of the cultural practices is to provide the crop with an edge against weeds in terms of competition. The row spacing was decreased so that the crop could reach the canopy more rapidly and so shade out the weeds. By rotating crops, you may prevent weeds from becoming used to the methods of weed management that are typically used for a single crop. The control of nutrients in order to minimize the amount of nutrients that are taken up by weeds while maximizing crop absorption. In order to compete with weeds for space, sunshine, nutrients, and water, cover crops allow for competition. Changed planting dates in order to provide the crop with a head start or to provide room for a flush of weed germination that may be suppressed prior to planting. The selection of crop varieties to guarantee that crops have the greatest possible advantage against weeds in terms of competition.

Chemical Methods

Even in integrated weed management (IWM) programs, herbicides will continue to play a similar role as they do in the majority of weed control strategies. Both accurate identification of weeds and understanding of the herbicide-resistant weeds that are present in the region are essential. Correct application of herbicides refers to the application of the appropriate substance at the appropriate rate and at the appropriate time respectively. Through the use of tank mixes that include herbicides that have several effective sites of action (SOA) and by rotating herbicides throughout the season whenever it is feasible to do so, enhanced variety was achieved. If you want to avoid applying herbicides with the same SOA over and over again, you should plan ahead for each season.

Herbicide that has been approved when it comes to herbicides, organic farmers are restricted to using just a select few natural compounds. Gluten-based corn meal: A by-product of the manufacturing of maize starch, corn gluten meal is the product that is used the most often in the United States of America. According to Finney and Creamer (2008), maize gluten meal has the potential to be used as a pre-emergence herbicide. The presence of gluten is necessary for the germination of weed seeds in order to prevent the development of roots; thus, the time of the application is very important. Corn gluten meal (Digitaria ischaemum) negatively affects a number of weeds, including redroot pigweed, black nightshade (Solanum nigrum), common lamb squatters, curly dock, creeping bent grass (Agrostis palustris), purslane, common dandelion (Taraxacumofficinale), and smooth crabgrass. These are just some of the weeds that are affected. Barnyard grass (Echinochloa crus-galli) and velvetleaf (Abutilon theophrasti) are the weeds that are the least vulnerable to the effects of maize gluten meal, according to Bingaman and Christians (1995). In general, grasses are less susceptible to the effects of corn gluten meal than plant species that have wide leaves. Corn gluten meal was applied before to planting, and the results of field trials showed that the amount of weed cover was decreased by as much as 84 percent (McDade and Christians, 2000).

Mechanical Methods

Mechanical methods of weed control involve the physical removal or disruption of weeds through various tools and machinery. This approach includes techniques such as tilling, mowing, hoeing, and hand-pulling. Tilling the soil can uproot weeds and bury their seeds, preventing germination, while mowing cuts down weeds before they can set seed, reducing their spread. Hoeing and hand-pulling are more labour-intensive but highly effective for small-scale operations or areas with heavy weed infestations. Mechanical weed control is advantageous because it reduces the need for chemical herbicides, thus lowering the risk of herbicide resistance and minimizing environmental impact. Additionally, these methods can be precisely targeted, making them suitable for both organic and conventional farming systems. By incorporating mechanical strategies, farmers can manage weed populations effectively while promoting soil health and sustainable agricultural practices.

Biological Methods

Biological methods of weed control leverage natural organisms to suppress weed populations, providing an environmentally friendly alternative to chemical herbicides. This approach utilizes various biological agents, including insects, fungi, bacteria, and grazing animals, which specifically target and reduce weed growth. For instance, certain insects feed on weed seeds or foliage, thereby limiting their ability to propagate. Similarly, pathogenic fungi and bacteria can infect and weaken weed species, reducing their competitive advantage. Grazing animals like goats or sheep can be strategically employed to consume and control weed infestations in pastures and fields. Biological control is advantageous as it integrates naturally into ecosystems, promoting biodiversity and reducing the environmental footprint of weed management. Additionally, it can be a sustainable solution for managing herbicide-resistant weed species. By employing biological methods, farmers can achieve effective and lasting weed control while enhancing ecological balance and soil health.

IWM in Rice

Effective weed control during the critical period is crucial, typically occurring between 20-30 days after treatment (DAT).

Approach based on culture

1) Manual weed removal
2) Manual weed extraction
3) Puddling
4) Dealing with Flooding

Integrated weed management (IWM) in rice cultivation involves a strategic combination of various control methods to effectively manage weed populations while promoting sustainable farming practices. This approach integrates cultural, mechanical, biological, and chemical tactics to create a comprehensive weed management plan. Cultural practices such as water management, crop rotation, and the use of competitive rice varieties help suppress weed growth by optimizing conditions unfavourable for weeds. Mechanical methods like manual weeding, the use of weeders, and precise tillage disrupt weed development and reduce seed banks in the soil. Biological controls involve the introduction of natural predators or the use of allopathic plants to inhibit weed growth. Chemical controls, when necessary, are applied selectively to target specific weeds without harming the rice crop or the environment. By combining these methods, IWM in rice ensures effective weed suppression, reduces the risk of herbicide resistance, and promotes ecological health, ultimately leading to higher yields and more sustainable rice production systems.

IWM in Wheat

Integrated weed management (IWM) in wheat cultivation employs a multifaceted approach to control weed populations, ensuring sustainable and productive farming. This strategy combines cultural, mechanical, biological, and chemical methods to create a robust weed management system. Cultural practices such as crop rotation, optimal planting times, and the use of competitive wheat varieties help suppress weeds by enhancing crop vigour and shading out potential invaders. Mechanical methods include tillage, harrowing, and manual weeding to physically remove or disrupt weed growth. Biological controls utilize natural predators, pathogens, or competitive cover crops to reduce weed prevalence. When necessary, chemical herbicides are applied judiciously and selectively to target specific weeds while minimizing environmental impact and the risk of resistance. By integrating these diverse techniques, IWM in wheat not only effectively manages weed populations but also supports soil health, biodiversity, and long-term agricultural sustainability, ultimately leading to improved wheat yields and quality.

IWM in Cotton

Integrated weed management (IWM) in cotton cultivation employs a holistic approach to effectively manage weed populations while promoting sustainable farming practices. This strategy combines cultural, mechanical, biological, and chemical methods to achieve optimal weed control. Cultural practices include crop rotation, cover cropping, and optimizing planting density to create an environment less favourable for weed growth. Mechanical methods involve cultivation, hoeing, and the use of specialized weeding equipment to physically remove weeds and disrupt their growth cycle. Biological controls incorporate the use of natural predators, competitive cover crops, and allelopathic plants to suppress weed proliferation. Chemical methods, when necessary, are used in a targeted and judicious manner to minimize herbicide resistance and environmental impact. By integrating these diverse tactics, IWM in cotton enhances weed suppression, promotes soil health, and supports biodiversity, leading to more resilient and productive cotton farming systems. This comprehensive approach not only ensures effective weed management but also contributes to the long-term sustainability and economic viability of cotton production.

IWM in Pulses

Integrated weed management (IWM) in pulse cultivation is essential for achieving sustainable and productive farming systems. Pulses, such as lentils, chickpeas, and beans, benefit significantly from IWM due to their sensitivity to weed competition. This approach combines cultural, mechanical, biological, and chemical methods to create an effective weed control strategy. Cultural practices, including crop rotation, intercropping, and using cover crops, help suppress weed growth by enhancing soil health and creating less favourable conditions for weeds. Mechanical methods, such as timely tillage and manual weeding, physically remove weeds and prevent their establishment. Biological control involves using natural predators or competitive crops to reduce weed pressure. Chemical methods are applied judiciously, targeting specific weed issues while minimizing herbicide use. By integrating these diverse strategies, IWM in pulses reduces the risk of herbicide resistance, promotes soil fertility, and supports overall ecosystem health. This comprehensive approach not only improves pulse yields and quality but also enhances the sustainability and economic viability of pulse farming.

IWM in Oil Seeds Crops

Integrated weed management (IWM) is crucial for optimizing oilseed crop production by effectively controlling weed populations while promoting sustainable agricultural practices. Oilseed crops, such as soybeans, sunflowers, and canola, are particularly vulnerable to weed competition, which can significantly reduce yields if left unmanaged. IWM integrates various control methods, including cultural, mechanical, biological, and chemical approaches, to combat weed infestations. Cultural practices like crop rotation, intercropping, and the use of cover crops help suppress weeds by enhancing soil health and biodiversity. Mechanical methods, such as tillage, mulching, and mechanical weeders, physically remove weeds and disrupt their growth. Biological controls leverage natural enemies of weeds or competitive crops to reduce weed pressure. Chemical controls, when necessary, are applied judiciously, targeting specific weed species while minimizing environmental impact. By combining these strategies, IWM in oilseed crops improves weed control efficacy, reduces herbicide dependency, and enhances overall crop productivity and quality. Moreover, IWM practices contribute to soil conservation, biodiversity preservation, and long-term sustainability in oilseed crop production systems.

Advantages of Integrated Weed Management

1. **Enhanced Weed Control**: By combining multiple control methods, IWM effectively targets various weed species and growth stages, resulting in more comprehensive weed suppression.

2. **Reduced Herbicide Dependence**: IWM minimizes the reliance on chemical herbicides, lowering the risk of developing herbicide-resistant weed populations and reducing chemical input costs.

3. **Environmental Protection**: With less chemical use, IWM mitigates the negative impact on soil, water, and non-target organisms, promoting a healthier ecosystem.

4. **Improved Crop Yields**: Effective weed management reduces competition for resources such as light, water, and nutrients, allowing crops to grow more vigorously and yield better.

5. **Soil Health Maintenance**: Practices like crop rotation and cover cropping within IWM enhance soil structure and fertility, contributing to long-term soil health.

6. **Biodiversity Support**: IWM encourages biodiversity by incorporating biological control agents and promoting diverse plant and animal life within agricultural systems.

7. **Economic Sustainability**: By reducing dependency on expensive herbicides and increasing crop productivity, IWM can improve the economic viability of farming operations.
8. **Adaptability and Flexibility**: IWM strategies can be tailored to specific crops, regions, and farming conditions, making it a versatile approach adaptable to various agricultural contexts.
9. **Resistance Management**: Diversifying weed control methods helps prevent or delay the development of resistance in weed populations to any single control tactic.
10. **Labour Efficiency**: While some methods may be labour-intensive initially, the overall integration of various tactics can lead to more efficient and manageable weed control systems over time.

Challenges in Integrated Weed Management

Integrated weed management (IWM) presents several challenges that need to be addressed to achieve effective and sustainable weed control. Key challenges include:

1. **Knowledge and Skill Requirements**: Implementing IWM requires a thorough understanding of various weed control methods and their interactions. Farmers need training and education to effectively integrate these practices.
2. **Increased Labour and Time**: Some IWM practices, like manual weeding and crop rotation, can be labour-intensive and time-consuming compared to conventional herbicide application.
3. **Initial Costs**: The adoption of IWM may involve higher initial costs for new equipment, seeds for cover crops, or biological control agents. These costs can be a barrier for some farmers.
4. **Complexity in Implementation**: Developing and maintaining a successful IWM program involves managing multiple components and adjusting strategies based on ongoing monitoring and environmental conditions, which can be complex and demanding.
5. **Variability in Effectiveness**: The effectiveness of IWM practices can vary based on local conditions, such as soil type, climate, and specific weed species present. What works well in one area may not be as effective in another.

6. **Herbicide Resistance Management**: While IWM aims to reduce herbicide reliance, the integration of chemical methods still requires careful management to prevent resistance. Misuse or overuse of herbicides within an IWM framework can still lead to resistance issues.
7. **Economic Constraints**: Although IWM can lead to long-term savings, the short-term economic benefits may not be immediately apparent, making it less attractive to farmers looking for quick returns.
8. **Regulatory and Policy Barriers**: In some regions, regulations and policies may not support or incentivize the adoption of IWM practices, limiting their widespread implementation.
9. **Availability of Resources**: Access to biological control agents, appropriate machinery, and alternative weed control materials can be limited, particularly in remote or resource-poor areas.
10. **Climate Change Impacts**: Changing climate conditions can alter weed dynamics, making it challenging to predict and manage weed populations effectively with existing IWM strategies.

Conclusion

Integrated weed management (IWM) represents a vital strategy for sustainable agriculture, combining multiple control methods to manage weed populations effectively while minimizing environmental impact. By integrating cultural, mechanical, biological, and chemical approaches, IWM enhances weed control efficacy, reduces herbicide reliance, and promotes long-term soil health and biodiversity. Despite challenges such as the need for increased knowledge, labour, and initial costs, the benefits of IWM—such as improved crop yields, resistance management, and economic sustainability—underscore its importance in modern farming. Continued research, education, and policy support are essential to overcoming implementation barriers and optimizing IWM practices. Ultimately, IWM offers a holistic solution that not only addresses current weed management issues but also contributes to the resilience and sustainability of agricultural systems.

References

Clëments, D. R., Weise, S. F., & Swanton, C. J. (1994). Integrated weed management and weed species diversity. Phytoprotection, 75(1), 1-18.

Harker, K. N., & O'Donovan, J. T. (2013). Recent weed control, weed management, and integrated weed management. Weed Technology, 27(1), 1-11.

Rao, N. A., & Nagamani, A. (2010). Integrated weed management in India-revisited.

Swanton, C. J., & Murphy, S. D. (1996). Weed science beyond the weeds: the role of integrated weed management (IWM) in agroecosystem health. Weed science, 44(2), 437-445.

Swanton, C. J., & Weise, S. F. (1991). Integrated weed management: the rationale and approach. Weed technology, 5(3), 657-663.

Swanton, C. J., Mahoney, K. J., Chandler, K., & Gulden, R. H. (2008). Integrated weed management: knowledge-based weed management systems. Weed Science, 56(1), 168-172.

Zhang, Z. P. (2003). Development of chemical weed control and integrated weed management in China. Weed Biology and Management, 3(4), 197-203.

13

Integrated Weed Management in Horticultural Crops

Anushi

Department of Fruit Science, CSAUAT, Kanpur, Uttar Pradesh

Abstract

Integrated weed management (IWM) is a critical aspect of sustainable horticultural crop production, offering a multifaceted approach to effectively control weed populations while minimizing environmental impact. Horticultural crops, including fruits, vegetables, and ornamentals, are highly susceptible to weed competition, which can reduce yields, harbour pests and diseases, and hinder crop quality. IWM integrates various control methods, including cultural, mechanical, biological, and chemical approaches, tailored to the specific needs and characteristics of horticultural systems. Cultural practices such as mulching, crop rotation, and intercropping help suppress weeds by creating unfavourable conditions for their growth and development. Mechanical methods, including hand-weeding, hoeing, and the use of specialized machinery, physically remove weeds and disrupt their growth cycle. Biological control strategies involve the introduction of natural predators or competitive plants to reduce weed populations. Chemical controls, when necessary, are applied selectively, targeting specific weed species while minimizing herbicide use and environmental impact. By combining these diverse tactics, IWM in horticultural crops improves weed control efficacy, enhances crop health and vigour, and promotes long-term sustainability in horticultural production systems. This holistic approach contributes to higher yields, better crop quality, and reduced reliance on chemical inputs, ultimately ensuring the economic viability and environmental sustainability of horticultural crop production.

Keywords: *crop rotation, chemicals, weeds, horticulture, mechanical*

Introduction

In reaction to both natural and forced surroundings, weed emerged and continues to evolve as an interfering association with our intended plant and activities. Weed is a plant that began in a natural environment and evolved as a result of these conditions. The plant that grows out of place and time is known as weed. They are undesirable and not helpful, tenacious and prolific, and they efficiently compete with the beneficial and desirable crop plants for space, nutrients, sunshine, and water. They also interfere with agricultural activities, which results in a decrease in the production and quality of food.

Because they interfere with the usage of land and water resources, weeds are plants that are not desired and do not deserve to be there. As a result, they have a negative impact on food output and the welfare of humans. As a result, a weed is a plant that is growing in an area where it is not desirable at the moment or that is not in its proper location. There are many different kinds of plants that are considered to be weeds. These include: sedges, grasses, broad-leafed weeds, aquatic plants, trees, and parasitic flowering plants (Striga, Orobanche). Weeds may have an impact on both agricultural areas and non-crop areas, such as industrial areas, road sides, railway lines, water tanks, irrigation channels, and so on.

Weeds are plants that are undesirable, unpleasant, and pernicious. They interfere with agricultural activities, drive up the amount of labour required, add to the expense of cultivation, and decrease the amount of crops that are harvested. There is no known species of weed; nonetheless, it has been proposed that it is a plant that is both dangerous and ineffective, and it grows persistently in areas where it is not needed.

The term "weed" refers to plants that are not needed or desired, that are growing in an inappropriate location, that impede the exploitation of natural resources, that are prolific, tenacious, competitive, dangerous, and even toxic in nature, and that are able to thrive in situations that are not favourable to their growth. Known as the "father of weed science," Jethro Tull.

Why Integrated Weed Management

There are certain circumstances in which a particular approach of weed management could be efficient and cost-effective, but in other circumstances, it might not be. There is not a single herbicide that is capable of properly managing a large variety of weed flora. The application of the same pesticide over and over again either produces a change in the flora or generates resistance in the weed flora that has escaped. For example, if you exclusively engage in one practice on a consistent basis, you can have some unfavourable impacts.

The Phalaris minor cropping method is used for rice and wheat. There is only one technique of weed management that has the potential to cause an increase in the population of a certain weed. The use of herbicides without discrimination and the consequences it has for both the environment and human health.

Cover Crops

Cover crops are annual or perennial plants that are cultivated between the rows of certain horticultural tree crops and vines. A grass sward is an example of a cover crop that may be either annual or perennial. Cover crops are utilized for a variety of reasons, and the management of these crops is not going to be covered in depth. The purpose of this discussion is to examine the ways in which they may be used for weed management, as well as the challenges, including weed control, that are involved with utilization of these methods. The nature of the business, the soil type, the rainfall pattern, and the danger of erosion.

It is possible to cultivate a wide range of cover crops, each of which is appropriate for the soil type and structure. Growing cover crops has the numerous advantages. The soil should be amended with organic materials, that is, either carbon or nitrogen. Enhance the traffic ability of machines for the purpose of suppressing the development of weeds. Species of legumes, grasses, and grains that are found in pastures, as well as a combination of other, less common volunteer species. There is the possibility that the cover crop may be permanent, such as green sods, or it may be eradicated and replanted annually. In order to get the most out of a cover crop, it is necessary to implement weed control measures in the areas where it is cultivated.

In dryland vine and tree crops, a permanent cover crop will compete for water with the crop; consequently, the growth of the cover crop is typically terminated in late spring or early summer. It is also important that cover crops be reduced in height prior to bud burst in spring, either through cultivation, mowing, or herbicide, in order to reduce the risk of frost. This is done in order to reduce the risk of frost.

Cover crops, if they are sufficiently competitive, have the potential to also serve as a type of weed management in their own right by pushing out species that are less desirable. In the following readings, a wide range of topics pertaining to the use of cover crops in horticultural tree crops are discussed. In order to get a comprehensive understanding of the benefits that cover crops provide, it is recommended that you read the first two readings, which are overviews of the use of cover crops in orchards. The following reading has further information about the drawbacks of cover crops, namely the issues that arise from the cover crop's excessive use of water, which may result in decreased yields.

Tillage in Soil

In all three farming methods, cultivation is still a common practice. Other means of weed management are gradually replacing cultivation, which was formerly much more commonly utilized. Planting vegetables on raised beds requires cultivation. Although less common, cultivation may also generate tree and vine crops. In addition, several rows still engage in cultivation. Incorporating a winter cover crop or maintaining a bare soil surface are two possible reasons to do this. Cultivation has negative effects on soil structure, yet it is nonetheless preferred by certain businesses since it allows them to eliminate all vegetation, which means less water competition. Both the disk plough and the rotary hoe are essential pieces of machinery for cultivating space between rows. Use a silly-plough or an under vine knife in vineyards to cultivate inside the row as well. Both viticulture and horticultural crops have these drawbacks. The degradation of soil caused by cultivation between rows, in particular, may make winter equipment manoeuvring problematic. Read on as Stirzaker (1991) delves into the challenges of reduced tillage in Australian vegetable production. Soil structure, biological activity, and crop yield are compared in the reading as they pertain to growing mulches and decreased tillage.

Herbicides

Integrated weed management (IWM) in horticultural crops often includes chemical control as one of its components. Chemical control involves the targeted use of herbicides to manage weed populations in horticultural systems. While chemical control is just one aspect of IWM and should be used judiciously and in conjunction with other control methods, it can be highly effective when applied correctly. Chemical control in horticultural crops offers several advantages. It provides quick and efficient weed control, especially in cases of severe weed infestations where manual or mechanical methods may be insufficient. Herbicides can also target specific weed species, minimizing damage to desirable crops and reducing competition for resources like water, nutrients, and sunlight. Additionally, chemical control can be essential for managing perennial or difficult-to-control weeds. However, there are also considerations and challenges associated with chemical control in horticultural crops within an IWM framework. Overreliance on herbicides can lead to the development of herbicide-resistant weed populations, which reduces the effectiveness of chemical control over time. Furthermore, herbicides can have unintended environmental impacts, such as soil and water contamination, as well as negative effects on non-target organisms.

To maximize the benefits of chemical control while minimizing its drawbacks, it should be integrated with other weed management practices within an IWM program. This may include cultural practices like mulching or crop rotation to reduce weed pressure, mechanical methods such as cultivation or hand-weeding for weed suppression, and biological controls like cover cropping or the use of natural enemies of weeds. By combining chemical control with these other strategies, horticultural growers can achieve effective weed management while promoting environmental sustainability and long-term crop health.

IWM in Mango

Integrated weed management (IWM) in mango cultivation is essential for maintaining orchard health, optimizing fruit yield and quality, and promoting long-term sustainability. Weeds compete with mango trees for vital resources such as water, nutrients, and sunlight, which can adversely affect tree growth and fruit production. Implementing IWM strategies tailored to mango orchards involves a combination of cultural, mechanical, biological, and chemical methods.

Cultural practices play a crucial role in IWM for mango cultivation. Mulching around the base of trees helps suppress weed growth by smothering weeds and conserving soil moisture. Intercropping with leguminous cover crops can provide additional weed suppression while fixing nitrogen in the soil. Proper spacing and pruning of mango trees also promote better light penetration and air circulation, reducing weed establishment.

Mechanical methods such as hand-weeding, hoeing, and mechanical weeders can be effective for controlling weeds in mango orchards, especially during the early stages of tree growth. Regular manual removal of weeds around the base of trees helps prevent competition for resources and reduces seed bank build-up.

Biological controls, such as using mulch from allelopathic plants or introducing weed-suppressive cover crops, can help inhibit weed growth while promoting beneficial soil microorganisms and biodiversity. Additionally, grazing animals like goats or sheep can be employed to manage weeds in orchards while providing additional income for farmers.

Chemical control, while often used as a last resort, can be necessary for managing persistent or invasive weed species in mango orchards. Herbicides should be selected carefully to minimize negative impacts on the environment, non-target organisms, and human health. Integrated with other control methods, chemical control can be an effective component of an overall IWM strategy for mango cultivation.

IWM in Grapes

When it comes to grape production, integrated weed management (IWM) is very necessary for ensuring the ongoing health of vineyards, maximizing grape output and quality, and fostering environmentally responsible viticulture practices. It is possible for weeds to have a detrimental influence on the growth of grapevines, the development of fruit, and the quality of wine because they compete with grapevines for critical resources such as water, nutrients, and sunshine. Methods that are cultural, mechanical, biological, and chemical are used in conjunction with one another in order to implement IWM tactics that are specifically customized to grapevine growth. In vineyards, cultural practices are essential components of integrated water management (IWM). It is possible to reduce the development of weeds, improve the structure of the soil, and increase biodiversity by planting cover crops between rows. The use of mulch around grapevines helps to preserve the moisture content of the soil, prevent the growth of weeds, and provide organic matter to the soil. The right spacing of vines, trellising, and management of the canopy all contribute to improved light exposure and airflow, which in turn reduces the development of weeds and the amount of competition they face. Weeds may be efficiently controlled by the use of mechanical means such as mowing, discing, or cultivating between vine rows. This is particularly true during the early phases of vine development. Both mechanical weeders and manual tools may be used for the purpose of achieving precise weed management in the vicinity of individual grapevines. During the process of mechanical weed control, however, it is essential to exercise caution so as not to do damage to the roots of the vines or to cause soil erosion. Controlling the development of weeds and promoting the presence of beneficial soil microbes and biodiversity in vineyards may be accomplished by the use of biological controls. These controls include the utilization of cover crops that possess allelopathic qualities or the introduction of plant species that inhibit weed growth. Additionally, grazing animals such as sheep or goats may be deployed to reduce weeds while also providing extra soil fertility via the deposition of manure.

IWM in Apples

When it comes to preserving the health of apple orchards and maximizing their yield, integrated weed management (IWM) is very necessary. Apple trees are in competition with weeds for vital resources such as water, nutrients, and sunshine, which may result in a reduction in the development of the trees and the amount of fruit they produce. Apple orchards need a variety of approaches, including cultural, mechanical, biological, and chemical ones, in order to successfully implement integrated water management (IWM) techniques. In apple orchards, cultural practices are essential elements that make up integrated winemaking (IWM). The use of mulch at the base of apple trees helps to prevent the development of weeds, preserve the moisture in the soil, and enhance the overall health of the soil. As a result of proper spacing between trees and frequent trimming, air circulation and light penetration are improved, which in turn reduces the formation of weeds and the amount of competition between them. It is possible to effectively manage weeds in apple orchards by the use of mechanical means such as mowing, cultivation, and hand-weeding, particularly during the early phases of tree development by using these approaches. Materials that are used for mulching, such as wood chips or straw, may also be used to suffocate weeds and prevent their seeds from germinating. In apple orchards, the application of biological controls, such as the establishment of cover crops or the intercropping of allelopathic plants, may be of assistance in reducing the development of weeds and enhancing the fertility of the soil. In addition, the introduction of valuable insects or the use of grazing animals such as sheep or goats may assist in the management of weeds while simultaneously fostering biodiversity.

Advantages of Integrated Weed Management

1. **Effective Weed Control**: IWM combines multiple control methods, such as cultural, mechanical, biological, and chemical approaches, to target weeds at different growth stages and with various characteristics, resulting in more comprehensive and effective weed suppression.

2. **Reduced Herbicide Dependency**: By integrating non-chemical weed control methods, IWM reduces reliance on herbicides, minimizing the risk of herbicide resistance development in weed populations and decreasing chemical input costs.

3. **Environmental Sustainability**: IWM promotes sustainable farming practices by minimizing the environmental impact associated with herbicide use, such as soil and water contamination, and preserving beneficial soil microorganisms and biodiversity.

4. **Improved Crop Yields and Quality**: Effective weed management through IWM reduces competition for resources like water, nutrients, and sunlight, allowing horticultural crops to grow more vigorously and produce higher yields of better quality fruits, vegetables, or flowers.

5. **Enhanced Soil Health**: Cultural practices within IWM, such as cover cropping and mulching, improve soil structure, fertility, and moisture retention, promoting long-term soil health and productivity.

6. **Biodiversity Support**: IWM encourages biodiversity by incorporating biological control agents and promoting diverse plant and animal life within horticultural systems, contributing to ecosystem resilience and stability.
7. **Economic Viability**: While initial investment in IWM practices may be required, the long-term benefits, including reduced input costs, increased yields, and improved crop quality, contribute to the economic sustainability of horticultural operations.
8. **Resistance Management**: By diversifying weed control methods, IWM helps prevent or delay the development of resistance in weed populations to any single control tactic, ensuring continued effectiveness of weed management strategies.
9. **Flexibility and Adaptability**: IWM practices can be tailored to specific crops, regions, and farming conditions, offering flexibility and adaptability to diverse horticultural systems and production environments.
10. **Labour Efficiency**: While some IWM practices may require initial labour input, the overall integration of various tactics can lead to more efficient and manageable weed control systems over time, reducing labour requirements and costs.

Conclusion

The conclusion is that integrated weed management, often known as IWM, is a method to weed control in horticulture crops that is both holistic and sustainable. By using a combination of cultural, mechanical, biological, and chemical techniques of weed control, integrated weed management (IWM) provides efficient weed suppression while simultaneously minimizing the impact on the environment, lowering the reliance on herbicides, and preserving ongoing soil health and biodiversity. Because of its many benefits, including increased crop yields and quality, better soil fertility, and economic viability, integrated water management (IWM) is becoming more important in contemporary horticultural production systems. Nevertheless, the implementation of IWM requires careful planning, continuous monitoring, and modification to the requirements of the local environment and the dynamics of the weeds. The advantages of integrated weed management (IWM) exceed the downsides, presenting a potential road towards sustainable weed control in horticulture. The problems that IWM faces include initial investment costs, the need for expertise and manpower, and the desire to have information. For the purpose of supporting the adoption and success of IWM methods, it is vital to continue research, education, and policy support. This will ensure the resilience and sustainability of horticultural crop production for future generations.

References

Merfield, C.N. (2023). Integrated weed management in organic farming. In Advances in Resting-state Functional MRI (pp. 31-109). Woodhead Publishing.

Nwilene, F. E., Nwanze, K. F., & Youdeowei, A. (2008). Impact of integrated pest management on food and horticultural crops in Africa. Entomologia experimentalis et applicata, 128(3), 355-363.

Robinson, D. E. (2014). Integrated Weed Management in horticultural crops. Recent advances in weed management, 239-254.

Singh, K. N., Singh, P., Singh, P., Alie, B. A., Bahar, F. A., & Panotra, N. (2010). Integrated weed management in vegetable crops under temperate conditions. Research Journal of Agricultural Sciences, 1(2), 160-163.

Singh, M., Kaul, A., Pandey, V., & Bimbraw, A. S. (2019). Weed management in vegetable crops to reduce the yield losses. International Journal of Current Microbiology and Applied Sciences, 8(7), 1241-1258.

Smeda, R. J., & Weston, L. A. (2017). Weed management systems for horticultural crops. In Handbook of weed management systems (pp. 553-601). Routledge.

14

Organic Crop Production for Sustainable Future

Anamika Pandey[1], Sreyansh Singh Chauhan[2], Alan Evon Singh[3] and Ajinkya Ajay Telgote[4]

[1,2,3,4]Department of Horticulture (Fruit Science), Naini Agricultural Institute SHUATS, Prayagraj, Uttar Pradesh, India

Abstract

Organic crop production is a cornerstone of sustainable agriculture, emphasizing environmental stewardship, biodiversity, and soil health. By eliminating synthetic pesticides and fertilizers, organic farming fosters resilient ecosystems and enhances soil fertility through natural processes such as composting and crop rotation. This method not only reduces pollution and greenhouse gas emissions but also promotes the conservation of water and energy resources. Organic crop production supports biodiversity by creating habitats for various species, contributing to a balanced ecosystem. Moreover, it offers health benefits by providing consumers with produce free from harmful chemicals. As global concerns about climate change and food security grow, organic farming represents a viable path towards a sustainable future, ensuring long-term agricultural productivity and ecological balance.

Introduction

Organic crop production is increasingly recognized as a critical component of sustainable agriculture. This approach to farming avoids the use of synthetic chemicals and genetically modified organisms (GMOs), relying instead on natural processes and materials to enhance soil fertility, control pests, and foster plant growth. The principles of organic farming are based on a holistic understanding of agricultural ecosystems, which prioritize environmental health, economic profitability, and social equity. This article explores the various aspects of organic crop production, its benefits, challenges, and its role in promoting a sustainable future.

Historical Context

The roots of organic farming can be traced back to traditional agricultural practices that predate industrialized agriculture. Early farmers used crop rotations, animal manure, and cover crops to maintain soil fertility and control pests. However, the advent of chemical fertilizers and pesticides in the early 20th century led to a dramatic shift in agricultural practices. The Green Revolution, which began in the 1940s, brought about increased food production through the use of high-yield crop varieties and synthetic inputs. While this revolution helped alleviate hunger in many parts of the world, it also led to environmental degradation, soil depletion, and a decline in biodiversity.

In response to these negative impacts, the organic farming movement began to take shape in the mid-20th century. Pioneers such as Sir Albert Howard, J.I. Rodale, and Lady Eve Balfour advocated for farming methods that worked in harmony with nature. The establishment of organizations such as the Soil Association in the UK and the Rodale Institute in the US provided a platform for research, education, and advocacy in organic farming.

Principles of Organic Farming

Organic farming is guided by a set of principles that emphasize the health of ecosystems, biodiversity, and sustainability. These principles provide a framework for practices that foster ecological balance and promote long-term agricultural productivity. The four core principles of organic farming are:

1. Health

Principle: Organic farming aims to sustain and enhance the health of soil, plants, animals, humans, and the planet as a whole.

Explanation

- **Soil Health**: Organic practices such as crop rotation, composting, and cover cropping improve soil structure, increase organic matter, and enhance microbial activity. Healthy soils lead to healthy crops that are more resilient to pests and diseases.
- **Plant Health**: By avoiding synthetic chemicals, organic farming reduces the risk of toxic residues in food, promoting better health for consumers.
- **Animal Health**: Organic livestock farming emphasizes humane treatment, including access to outdoor spaces, and prohibits the routine use of antibiotics and growth hormones.

- **Human Health**: Organic farming minimizes exposure to harmful chemicals for both farm workers and consumers, contributing to overall well-being.
- **Environmental Health**: Organic methods reduce pollution and protect natural resources, leading to healthier ecosystems.

2. Ecology

Principle: Organic farming should be based on living ecological systems and cycles, work with them, emulate them, and help sustain them.

Explanation

- **Ecosystem Integration**: Organic farming integrates agricultural practices with the natural environment, promoting biodiversity and ecological balance.
- **Nutrient Cycling**: Techniques like composting and green manuring recycle nutrients within the farm system, reducing the need for external inputs.
- **Water Conservation**: Organic practices such as mulching and cover cropping improve soil water retention and reduce runoff, conserving water resources.
- **Habitat Protection**: Organic farms often maintain natural habitats such as hedgerows, ponds, and woodlands, supporting wildlife and enhancing biodiversity.
- **Energy Efficiency**: Organic farming generally requires less energy than conventional farming due to the reduced reliance on synthetic inputs.

3. Fairness

Principle: Organic farming should build on relationships that ensure fairness with regard to the common environment and life opportunities.

Explanation

- **Social Equity**: Organic farming promotes fair wages and working conditions for farm workers, contributing to social justice.
- **Fair Trade**: Organic farming supports fair trading practices, ensuring that farmers receive fair prices for their products.
- **Animal Welfare**: Organic standards require humane treatment of livestock, providing them with conditions that meet their natural behaviors and needs.

- **Community Support**: Organic farming often emphasizes local production and distribution, supporting local economies and fostering community relationships.

4. Care

Principle: Organic farming should be managed in a precautionary and responsible manner to protect the health and well-being of current and future generations and the environment.

Explanation

- **Precautionary Principle**: Organic farming adopts a precautionary approach to new technologies and practices, ensuring they do not harm human health or the environment.
- **Sustainability**: Practices are designed to be sustainable in the long term, avoiding depletion of natural resources and promoting ecological balance.
- **Continuous Improvement**: Organic farming encourages continuous learning and improvement, adapting practices based on new knowledge and insights.
- **Environmental Stewardship**: Organic farmers are stewards of the land, taking responsibility for the environmental impacts of their practices and striving to minimize negative effects.

Benefits of Organic Crop Production

Environmental Benefits

1. **Soil Health**: Organic farming practices, such as crop rotation, cover cropping, and the use of organic fertilizers (compost, green manure), enhance soil structure and fertility. These practices increase organic matter in the soil, improving water retention and reducing erosion.
2. **Biodiversity**: Organic farms typically have higher biodiversity than conventional farms. The absence of synthetic pesticides and fertilizers creates a more hospitable environment for a wide range of organisms, including beneficial insects, birds, and soil microbes.
3. **Water Conservation**: Organic farming practices reduce water pollution by eliminating the runoff of synthetic chemicals into waterways. Techniques such as mulching and organic matter incorporation enhance soil water retention, reducing the need for irrigation.

4. **Climate Change Mitigation**: Organic farming can help mitigate climate change by sequestering carbon in the soil. Practices like cover cropping and reduced tillage increase soil organic carbon levels. Additionally, organic farming generally requires less energy input than conventional farming, reducing greenhouse gas emissions.

Economic Benefits

1. **Market Demand**: There is a growing consumer demand for organic products, driven by concerns about health, food safety, and environmental sustainability. This demand often translates into premium prices for organic produce, offering economic incentives for farmers.
2. **Farm Resilience**: Organic farms often exhibit greater resilience to environmental stressors such as droughts and floods. The emphasis on soil health and biodiversity can help buffer crops against extreme weather events and pest outbreaks, reducing financial risk for farmers.
3. **Job Creation**: Organic farming tends to be more labor-intensive than conventional farming, creating more jobs per unit of production. This can contribute to rural development and economic stability in farming communities.

Health Benefits

1. **Nutrient Content**: Some studies suggest that organic crops may contain higher levels of certain nutrients, such as antioxidants, than conventionally grown crops. While the differences are not always significant, they can contribute to a healthier diet.
2. **Reduced Exposure to Chemicals**: Organic farming reduces the exposure of farm workers and consumers to potentially harmful synthetic chemicals. This can lead to better health outcomes for both producers and consumers.
3. **Antibiotic Resistance**: Organic farming prohibits the routine use of antibiotics in livestock, which can help mitigate the spread of antibiotic-resistant bacteria, a significant public health concern.

Challenges of Organic Crop Production

Organic crop production faces several challenges that stem from the limitations of organic farming practices and the broader agricultural and socio-economic context. These challenges can affect farm productivity, profitability, and sustainability. Here are some of the key challenges:

1. **Pest and Disease Management**
   - **Limited Control Options**: Organic farmers cannot rely on synthetic pesticides, which may limit their ability to manage pests and diseases effectively.
   - **Resistance Development**: Pests and diseases may develop resistance to organic-approved pesticides, reducing their efficacy over time.
   - **Labor Intensity**: Organic pest and disease management often require more labor-intensive practices such as hand-weeding, crop rotation, and biological controls.
2. **Weed Control**
   - **Mechanical Methods**: Managing weeds without synthetic herbicides requires mechanical methods such as hand-weeding, cultivation, and mulching, which can be time-consuming and labor-intensive.
   - **Limited Efficacy**: Organic weed control methods may not be as effective as chemical herbicides, leading to higher weed pressure and reduced crop yields.
3. **Soil Fertility Management**
   - **Slow Release of Nutrients**: Organic fertilizers, such as compost and manure, release nutrients slowly over time, which can make it challenging to meet crop nutrient demands.
   - **Balancing Nutrient Inputs**: Organic farmers must carefully manage nutrient inputs to maintain soil fertility without overloading the soil with nutrients, which can lead to pollution of waterways.
4. **Market Access and Economics**
   - **Certification Costs**: Obtaining organic certification can be expensive, especially for small-scale farmers, due to inspection fees, paperwork, and compliance with standards.
   - **Price Volatility**: Organic crop prices can be volatile and may not always provide a sufficient premium over conventional prices to offset higher production costs.
   - **Transition Period**: Farmers transitioning from conventional to organic production face a period of reduced yields and increased costs during the transition period, which can be financially challenging.

5. Knowledge and Education

- **Technical Expertise**: Successful organic farming requires knowledge and skills in organic production techniques, pest and disease management, and soil fertility management, which may not be readily available to all farmers.
- **Access to Information**: Some farmers may lack access to education, training, and resources on organic farming practices, limiting their ability to adopt and implement organic methods effectively.

6. Policy and Institutional Support

- **Subsidies and Incentives**: Government agricultural policies and subsidies often favor conventional agriculture over organic farming, making it difficult for organic farmers to compete on a level playing field.
- **Research Funding**: Limited research funding and institutional support for organic agriculture may hinder the development of innovative organic farming practices and technologies.

7. Market Demand and Consumer Awareness

- **Consumer Education**: Despite growing interest in organic products, consumer awareness about the benefits of organic farming may be limited, leading to challenges in market development.
- **Market Infrastructure**: Some regions may lack the necessary infrastructure for organic production, processing, and distribution, limiting market access for organic farmers.

8. Climate Change and Environmental Factors

- **Weather Variability**: Climate change can lead to increased weather variability, including droughts, floods, and extreme temperatures, which can impact crop yields and production stability.
- **Pest and Disease Pressure**: Changes in climate patterns may also alter pest and disease pressure, requiring organic farmers to adapt their pest management strategies accordingly.

Organic Crop Production Techniques

1. Soil Management

a. **Crop Rotation**: Rotating crops on a regular basis helps break pest and disease cycles, improves soil structure, and maintains fertility. Different crops have different nutrient needs, so rotation also prevents nutrient depletion.

b. **Cover Cropping**: Planting cover crops, such as legumes or grasses, during fallow periods helps prevent soil erosion, suppress weeds, and improve soil fertility by adding organic matter when the cover crops are incorporated into the soil.

c. **Green Manure**: Similar to cover cropping, green manure involves growing specific crops that are later incorporated into the soil to improve fertility. Leguminous crops are often used because they fix nitrogen from the air into the soil.

d. **Composting**: Recycling organic materials like crop residues, kitchen scraps, and animal manure into compost enriches the soil with nutrients, improves soil structure, and enhances microbial activity.

2. Weed Management

a. **Mechanical Cultivation**: Hand weeding, hoeing, or using mechanical tools like cultivators help control weeds without relying on herbicides.

b. **Mulching**: Applying organic mulches, such as straw, hay, or wood chips, around plants suppresses weed growth, conserves soil moisture, and adds organic matter to the soil as the mulch decomposes.

c. **Flame Weeding**: Using propane-fueled flame weeders to pass over weeds quickly kills them by heat without harming the soil or crops.

3. Pest and Disease Management

a. **Biological Control**: Introducing or encouraging natural predators, parasites, and beneficial organisms to control pest populations. This can include releasing ladybugs to eat aphids or using parasitic wasps to control caterpillars.

b. **Crop Diversity**: Planting a variety of crops and creating habitat for beneficial insects increases biodiversity and reduces the likelihood of pest outbreaks.

c. **Crop Resilience**: Choosing pest-resistant crop varieties adapted to local conditions helps reduce susceptibility to pests and diseases.

4. Nutrient Management

a. **Organic Fertilizers**: Using natural fertilizers such as compost, manure, bone meal, and seaweed extracts provides essential nutrients to crops while improving soil fertility and structure.

b. **Vermicomposting**: Composting with earthworms produces nutrient-rich vermicompost, which can be added to soil to enhance fertility and microbial activity.

c. **Mineral Amendments**: Applying natural minerals like rock phosphate, limestone, or gypsum can adjust soil pH and provide specific nutrients that may be lacking.

5. Water Management

a. **Drip Irrigation**: Using drip irrigation systems delivers water directly to the base of plants, minimizing water waste and reducing weed growth compared to overhead irrigation.

b. **Rainwater Harvesting**: Collecting rainwater in barrels or cisterns for later use reduces dependence on municipal water sources and conserves water.

c. **Soil Moisture Conservation**: Practices such as mulching, cover cropping, and reduced tillage help retain soil moisture and reduce evaporation.

6. Genetic Selection and Plant Breeding

a. **Heirloom Varieties**: Choosing heirloom or open-pollinated varieties adapted to local growing conditions can improve crop resilience and reduce the need for chemical inputs.

b. **Participatory Plant Breeding**: Involving farmers in the breeding process helps develop crop varieties suited to organic farming systems and local environments.

7. Integrated Pest Management (IPM)

a. **Monitoring and Scouting**: Regularly inspecting crops for signs of pests and diseases allows for early detection and intervention.

b. **Threshold Levels**: Setting action thresholds based on pest population levels helps determine when control measures are necessary.

c. **Cultural Controls**: Practices like crop rotation, sanitation, and habitat manipulation disrupt pest life cycles and reduce reliance on chemical controls.

Case Studies and Success Stories

1. Rodale Institute's Farming Systems Trial

Location: Kutztown, Pennsylvania, United States

Description: The Rodale Institute, a pioneering organization in organic farming research, initiated the Farming Systems Trial (FST) in 1981 to compare the long-term effects of organic and conventional farming systems. The FST is the longest-running side-by-side comparison of organic and conventional farming systems in North America. Results from the trial have shown that organic farming methods can match or exceed the yields of conventional agriculture, especially during drought years, while improving soil health, biodiversity, and nutrient retention.

2. Sikkim: India's First Organic State

Location: Sikkim, India

Description: In 2016, Sikkim became the first state in India to be certified as fully organic under the "Organic Mission." The government of Sikkim implemented a phased approach to transition the state's agriculture from conventional to organic practices. Farmers were provided with training, technical support, and financial incentives to adopt organic methods. The transition to organic farming has led to improvements in soil health, water quality, and biodiversity while boosting tourism and creating economic opportunities for farmers.

3. Blaencamel Farm: Agroecological Farming in Wales

Location: Aberystwyth, Wales, United Kingdom

Description: Blaencamel Farm is a family-run farm in Wales that practices agroecological farming methods. The farm produces a diverse range of organic vegetables, herbs, and flowers using techniques such as crop rotation, companion planting, and minimal tillage. By prioritizing soil health and biodiversity, Blaencamel Farm has been able to increase yields, reduce pest and disease pressure, and improve the resilience of their crops to environmental stressors. The farm also sells its produce directly to consumers through farmer's markets and community-supported agriculture (CSA) schemes, fostering connections between farmers and consumers.

4. La Durette: Biodynamic Vineyard in France

Location: La Durette, Provence, France

Description: La Durette is a biodynamic vineyard located in the Provence region of France. Biodynamic farming goes beyond organic practices by incorporating spiritual and holistic principles into agriculture. La Durette follows biodynamic principles such as planting according to lunar cycles, using herbal preparations to enhance soil fertility, and fostering biodiversity in the vineyard. These practices have led to healthier vines, improved grape quality, and a unique terroir expression in the wines produced at La Durette. The vineyard has garnered acclaim for its high-quality, biodynamically grown wines and serves as an example of sustainable viticulture.

5. Grameen Danone Foods: Socially Responsible Organic Dairy in Bangladesh

Location: Bogura, Bangladesh

Description: Grameen Danone Foods is a joint venture between Grameen Bank and Danone that produces fortified yogurt products in Bangladesh. The company sources organic milk from local smallholder farmers who practice sustainable dairy farming methods. Grameen Danone provides training, support, and fair prices to dairy farmers, empowering rural communities and improving livelihoods. The company's commitment to organic and socially responsible practices has helped address malnutrition and food insecurity in rural Bangladesh while creating economic opportunities

The Future of Organic Crop Production

Technological Innovations

Advances in technology offer new opportunities for enhancing organic farming practices. Innovations such as precision agriculture, which uses GPS and data analytics to optimize farming practices, can improve efficiency and reduce labour costs. Additionally, the development of organic biopesticides and biofertilizers can provide more effective and environmentally friendly alternatives to synthetic chemicals.

Policy and Institutional Support

Government policies and institutional support are crucial for the expansion of organic farming. Policies that provide financial incentives, research funding, and infrastructure development can help farmers transition to and sustain organic practices. The establishment of organic certification programs and standards is also essential for maintaining consumer trust and market integrity.

Consumer Education and Awareness

Increasing consumer awareness about the benefits of organic farming is key to driving demand for organic products. Educational campaigns, labeling initiatives, and transparent communication about organic practices can help consumers make informed choices and support sustainable agriculture.

Global Collaboration

Addressing the global challenges of food security, climate change, and environmental degradation requires international collaboration. Sharing knowledge, resources, and best practices across countries and regions can enhance the effectiveness of organic farming and contribute to a more sustainable global food system.

Conclusion

Organic crop production represents a viable and sustainable alternative to conventional farming. By prioritizing soil health, biodiversity, and ecological balance, organic farming can contribute to a more resilient and sustainable agricultural system. While challenges remain, the growing demand for organic products, advancements in technology, and supportive policies offer promising opportunities for the future. As the world grapples with the pressing issues of climate change, food security, and environmental degradation, organic farming provides a pathway towards a more sustainable and equitable future.

References

Aher, S. B., Bhaveshananda, S., & Sengupta, B. (2012). Organic agriculture: Way towards sustainable development. International Journal of Environmental Sciences, 3(1), 209-216.

Azadi, H., Schoonbeek, S., Mahmoudi, H., Derudder, B., De Maeyer, P., & Witlox, F. (2011). Organic agriculture and sustainable food production system: Main potentials. Agriculture Ecosystems & Environment, 144(1), 92-94.

Brzozowski, L., & Mazourek, M. (2018). A sustainable agricultural future relies on the transition to organic agroecological pest management. Sustainability, 10(6), 2023.

Hans, V. B., & Rao, R. (2018). Organic farming for sustainable development in India. Acta Scientific Agriculture, 2(12), 96-102.

Kilcher, L. (2007). How organic agriculture contributes to sustainable development. JARTS Witzenhausen, Supplement 89, 31-49.

Niggli, U. (2015). Sustainability of organic food production: challenges and innovations. Proceedings of the Nutrition Society, 74(1), 83-88.

Pasupulla, A. P., Pallathadka, H., Nomani, M. Z. M., Salahuddin, G., & Rauf, M. (2021). A survey on challenges in organic agricultural practices for sustainable crop production. Annals of the Romanian Society for Cell Biology, 338-347.

15

Integrated Water Management in Agricultural Crops

Ritika Raj

Department of Agronomy, Banaras Hindu University, Varanasi Uttar Pradesh

Abstract

The method known as integrated water management (IWM) in agricultural crops is an all-encompassing strategy that aims to maximize the effectiveness of water use, enhance crop output, and guarantee the management of water resources in a sustainable manner. This method incorporates a number of different water management strategies, such as the timing of irrigation, the use of water-saving devices, the monitoring of soil moisture, and conservation measures. Combining these methods allows IWM to achieve its goal of limiting water losses caused by evaporation, runoff, and deep percolation while simultaneously ensuring that crop water demand is met by the available water supply. Mulching, cover cropping, and crop rotation are examples of cultural techniques that assist improve soil moisture conservation and decrease water loss via evaporation. The application of water may be made more exact and targeted by the use of mechanical technologies like as precision irrigation systems, drip irrigation, and soil moisture sensors. This helps to maximize the efficiency with which crops utilize water. Improvements in water retention and plant resistance to water stress may be achieved via the use of biological controls. These controls include the selection of drought-resistant crop cultivars and the enhancement of the organic matter content of the soil. Chemical controls, such as the use of soil conditioners and wetting agents, have the potential to improve soil structure and infiltration, hence permitting improved water distribution and absorption by crops. In general, integrated water management (IWM) plays a significant part in reducing the issues associated with water shortages, fortifying agricultural systems against the effects of climate change, and fostering sustainable water use practices in agricultural systems.

Keywords: *application, evaporation, percolation, maximize*

Introduction

In addition to being the source of life on earth, water is also an essential component of the overall 2030 Agenda (UN-Water, 2016). Every single one of the Sustainable Development Goals (SDGs) is dependent on water. For the purpose of fulfilling the 2030 Agenda for Sustainable Development, it is vital to address the global water crisis, which includes water shortages, drought, floods, and water pollution. In order to value and manage freshwater in an integrated way, quick action, strong political will, and greater finance are required. Water is essential to our well-being and survival. The use of freshwater from rivers, lakes, and aquifers is essential to agricultural production. A portion of livestock production and agriculture that is dependent on rainwater are dependent on the water that is provided by limited and changeable rainfall patterns. Furthermore, freshwater ecosystems enable people to maintain their livelihoods, ensure their food security, and maintain their nutrition by, among other things, sustaining inland fisheries and aquaculture. There is a need for supplies of freshwater that is free of contamination in order to assure the safety of drinking water, as well as to guarantee that hygiene and food safety regulations are met. In addition, water is used for a wide variety of additional purposes and is essential to the activities of humans. According to forecasts made on a global scale, the demand for freshwater is expected to dramatically expand over the next several decades as a result of population expansion, dietary diversity, economic development, and urbanization.

At the same time, recent evaluations, predictions, and scenarios from the worldwide community offer an alarming picture of the planet's water resources. These assessments stress the overuse, abuse, degradation, pollution, and rising scarcity of water, as well as the hazards of drought and floods that are associated with climate change. Flooding is often associated with or provokes other types of natural catastrophes, such as landslides. Floods not only have an effect on individuals who are immediately impacted by them, but they may also have wider-reaching consequences that can cascade (and often intensify) via related infrastructure networks and supply chains (affecting essential services such as food supplies, electricity, and communications), social isolation, and stress. It is possible that recuperation will take a significant amount of time after the flood has subsided, as people, towns, and countries work to reconstruct their lives and economy. Therefore, when it comes to tackling flood risks and the cascading effects of those risks, it is necessary to use a multi-hazard catastrophe risk reduction strategy. Effective agricultural flood risk management, which takes into account the importance of food security, has the potential to play a crucial role in supporting the results that are sought in terms of society, the environment, and the economy. In light of this, and in contrast to the paradigm of flood control, which is frequently characterized by a narrowly defined single objective, strategic flood risk management places an emphasis on reducing risk (to people, economies, and the environment) and on building resilience. It also seeks opportunities to work with natural processes and promote multiple benefits across a variety of sectors, such as crops and livestock, energy, fisheries, societal well-being, and the environment.

There is a close relationship between water, food, climate, and energy, with agriculture being a significant participant in the process of climate change. Agriculture is responsible for 70% of the world's water use and 6% of the world's energy consumption. Irrigated agriculture makes a substantial contribution to the production of food on a worldwide scale, as well as to the security of food supplies and the incomes of rural residents. Rainfed grain yields are expected to be 2.7 tons per hectare, whereas irrigated cereal yields are estimated to be 4.4 tons per hectare. In the absence of irrigation, the output of cereals throughout the world would be 25% lower.

From now until 2050, it is anticipated that the demand for food will rise by 70–90% as a result of changes in diets and the expansion of the population. It is anticipated that the world population would reach its highest point of 9.2 billion in the year 2050, with the majority of the expansion happening in developing and least developed nations. As more people move into urban areas, their eating patterns change from being centered on cereal to being based on meat. In most cases, the cultivation of livestock requires a greater quantity of water than the production of grains. If there is no increase in water productivity, the need for water specifically for food production will treble over the next several decades.

Due to the fact that 240 million people, or thirty percent of the population, in sub-Saharan Africa are food insecure, food security is a priority for governments, donors, and non-governmental organizations. People's capacity to get food, the amount of food available, and the quality of the food all play a role in determining food security. It is possible to describe this by referring to the three A's, which are accessibility, accessibility, and availability.

Methods to Conserve Water

Both irrigation and drainage play an important part in the supply of ecosystem services, which include the production of food, the delivery of services such as fish and wood, and the regulation of ecosystem services such as the recharging of groundwater, the retention of floodwater and sediment, and the removal of carbon from the atmosphere. The control of erosion, the build-up of organic matter in the soil, the recycling of soil nutrients, and the reinforcement of species variety via the provision of habitats for flora and animals are all examples of ecosystem services that are beneficial.

A key problem that arises in many agricultural regions as a result of excessive pumping is the recharging of groundwater. A significant contributor to the replenishment of groundwater and the preservation of biodiversity is seepage from canals, reservoirs, agricultural ponds, and irrigation fields. For example, farm ponds in Madhya Pradesh collect and store runoff for the purpose of irrigating crops. This helps to contribute to the recharging of groundwater and the preservation of biodiversity. The seepage that occurs from rice fields in Taiwan is responsible for forty percent of the groundwater recharge, and it has an economic value that is roughly sixty percent of the paddy output.

Another essential component of irrigation is flood retention, which is a ccomplished via the use of flood retention basins. Because of their ability to hold water and the combined impact of seepage and evaporation from bunded fields, irrigatedpaddyfieldshavethepotentialtoplayasubstantialroleinfloodprevention. This is because the runoff from these fields is much lower than that received from rainfed fields. The prevention of floods is much improved as a result of this. Because soil erosion is one of the most significant contributors to soil loss and degradation, terracing is an efficient method for halting or minimizing the impact of soil erosion. It is possible to increase agricultural output while simultaneously providing advantages to other ecosystems by terracing irrigated paddy fields in hilly places. Terracing minimizes the amount of tillage conducted, restores land cover, and maintains water retention in the soil. The agricultural sector is the most significant contributor to climate change, both directly via emissions of greenhouse gases (GHG) and indirectly through changes in land use alterations. Agriculture is the sector with the greatest potential for climate change mitigation, notably via the storage of carbon in soils. There is the possibility that another option that might help to carbon sequestration is the irrigation of dry and semiarid areas.

The production of methane from paddy fields is a contributor to the phenomenon of climate change. There are a variety of alternative water management strategies that may help minimize methane emissions. Some examples of these practices are alternating wet-dry irrigation and aeration via mid-season drainage. There is a significant contribution to carbon dioxide emissions made by the oxidation of peatlands that have been drained and converted to agricultural land. A balance between high groundwater levels, which may minimize greenhouse gas emissions, and low groundwater levels, which allow for farming, can be achieved by improved drainage methods.

Ensuring the preservation of biodiversity is yet another essential component of irrigation and drainage. Over five thousand species have been documented in irrigated paddy fields in Japan, and the cultivation of rice in a manner that is mindful of biodiversity has been encouraged via participation in biodiversity certification programs and cooperation between many stakeholders.

Precision Irrigation

Precision irrigation is a cutting-edge approach to water management in agriculture that aims to optimize water use efficiency and crop productivity while minimizing water waste. Unlike traditional irrigation methods, which often involve uniform water application across entire fields regardless of crop water needs, precision irrigation delivers water precisely and efficiently to specific areas or plants within a field. This is achieved through the use of advanced technologies such as drip irrigation, micro-irrigation, and soil moisture sensors.

Drip irrigation systems deliver water directly to the root zone of plants through a network of tubes and emitters, minimizing evaporation and runoff. Micro-irrigation systems, including micro-sprinklers and micro-jet systems, apply water in small, localized amounts, reducing water wastage and ensuring precise water delivery to individual plants. Soil moisture sensors are used to monitor soil moisture levels in real-time, allowing farmers to tailor irrigation schedules and amounts based on actual crop water needs.

Precision irrigation offers several benefits to farmers and the environment. By providing crops with the right amount of water at the right time, it helps optimize crop growth and yield while conserving water resources. Additionally, precision irrigation can improve nutrient uptake and reduce the leaching of fertilizers and pesticides into the environment. Overall, precision irrigation plays a crucial role in sustainable agriculture by promoting water efficiency, environmental stewardship, and long-term agricultural viability.

Examples of crops

- **Utilizing Drip Irrigation for Maize:** When compared to the more conventional flood irrigation techniques, the use of drip irrigation systems in maize production may dramatically lower the amount of water that is used. Through the use of drip irrigation, water is delivered directly to the root zone of maize plants, therefore reducing the amount of water that is lost to evaporation and runoff while simultaneously increasing the amount of water that is taken up by the plants.

- **Wheat Production and Conservation Tillage:** The use of conservation tillage measures, such as reduced tillage or no-till, in wheat production helps to enhance the structure of the soil and the amount of moisture that is retained. Through the practice of leaving crop residues on the surface of the soil, conservation tillage helps to prevent soil erosion and water runoff, hence increasing the amount of water that is available for wheat crops.
- **Cover Cropping with Soybeans:** The incorporation of cover crops such as clover or ryegrass into soybean rotations facilitates the protection of the soil surface, the reduction of evaporation, and the enhancement of the soil's moisture content. In addition, cover crops provide organic matter to the soil, which elevates the soil's ability to retain water and increases its fertility, which is beneficial to following soyabean crops.
- In cotton fields, the use of mulch, whether it be organic or artificial, helps to maintain soil moisture, limit the development of weeds, and lessen changes in soil temperature. Mulching also minimizes the amount of water that evaporates from the surface of the soil, which enables cotton plants to receive water more effectively and enables them to stay at their ideal development rate.
- Monitoring Soil Moisture with Rice Farmers are able to more properly plan irrigation and monitor the levels of soil moisture in rice fields by installing soil moisture sensors in the paddies. Farmers are able to maximize the use of water and reduce the quantity of water that is wasted in rice production by modifying the timing and amount of irrigation based on real-time data about the moisture content of the soil.
- Rainwater harvesting systems may be used in barley fields to augment irrigation water sources and minimize dependency on groundwater or surface water. This can be accomplished via the combination of rainwater gathering with barley. Rainwater that has been collected may be kept in reservoirs or ponds and then utilized to irrigate barley fields during dry seasons. This increases the amount of water that is available and the amount of crop that can be produced.

Advantages

1. **Optimized Water Use Efficiency**: IWM enables farmers to maximize water use efficiency by matching water supply with crop water demand, reducing water wastage through evaporation, runoff, and deep percolation.

2. **Improved Crop Productivity**: By providing crops with the right amount of water at the right time, IWM helps optimize crop growth, development, and yield, leading to improved agricultural productivity and profitability.
3. **Sustainable Water Resource Management**: IWM promotes the sustainable use and conservation of water resources by minimizing water extraction from freshwater sources, reducing reliance on groundwater pumping, and enhancing water reuse and recycling.
4. **Mitigation of Water Scarcity**: In regions prone to water scarcity or drought, IWM practices can help mitigate the impacts of water shortages by improving water availability for agricultural production and maintaining crop yields under limited water conditions.
5. **Environmental Protection**: By reducing water runoff, erosion, and leaching of agrochemicals into water bodies, IWM practices contribute to the protection of aquatic ecosystems, biodiversity, and water quality.
6. **Climate Change Adaptation**: IWM enhances agricultural resilience to climate change by improving soil moisture management, reducing vulnerability to water stress, and mitigating the impacts of extreme weather events on crop production.
7. **Cost Savings**: Implementing IWM practices can lead to cost savings for farmers by reducing water pumping and irrigation costs, minimizing input costs associated with fertilizers and pesticides, and enhancing overall farm profitability.
8. **Long-term Sustainability**: By integrating multiple water management strategies, IWM promotes long-term sustainability in agriculture, ensuring the continued viability of agricultural production systems while safeguarding water resources for future generations.

Challenges

1. **Limited Awareness and Education**: Many farmers may lack awareness of IWM practices or have limited access to training and education on water management techniques. Increasing awareness and providing extension services are essential for promoting the adoption of IWM.
2. **High Initial Investment Costs**: Implementing IWM practices often requires significant upfront investment in infrastructure, technology, and training. High costs can be a barrier for farmers, particularly those in resource-constrained settings or smallholder farmers.

3. **Technical Complexity**: Some IWM practices, such as precision irrigation systems or soil moisture monitoring, may be technically complex and require specialized knowledge for proper implementation and maintenance. Farmers may need support and training to adopt these practices effectively.
4. **Access to Resources**: Limited access to water-saving technologies, irrigation infrastructure, and financial resources can hinder the adoption of IWM, particularly in rural and marginalized communities.
5. **Policy and Institutional Barriers**: Inadequate policies, regulations, and institutional support may hinder the adoption of IWM practices. Policies that promote water conservation, incentivize sustainable water use, and provide financial support for IWM implementation are essential for overcoming these barriers.
6. **Social and Cultural Factors**: Socio-cultural factors, such as traditional farming practices and social norms, can influence farmers' willingness to adopt new water management techniques. Understanding local contexts and engaging with communities are crucial for successful IWM implementation.
7. **Water Governance and Management**: Inefficient water governance structures, conflicts over water allocation, and competing water demands from various sectors can pose challenges for implementing IWM at the basin or watershed level. Coordinated water management approaches and stakeholder engagement are necessary for addressing these challenges.
8. **Climate Change Impacts**: Climate change exacerbates water scarcity and variability, posing additional challenges for water management in agriculture. Adapting IWM practices to changing climatic conditions and building resilience to climate-related risks are critical for sustainable water management.

Conclusion

In conclusion, integrated water management (IWM) has a number of issues that need to be solved, despite the fact that it offers a great deal of advantages for the management of water resources and sustainable agriculture. Limited awareness and education, high initial investment costs, technical complexity, availability to resources, legislative and institutional impediments, social and cultural considerations, water governance difficulties, and the implications of climate change are some of the challenges that are now being faced. The implementation of IWM methods has the potential to significantly improve water usage efficiency, boost agricultural output, and guarantee water security for both communities and farmers, despite the challenges that have been presented. In order to overcome these problems and fully achieve the advantages of integrated water management (IWM), it is necessary for governments, policymakers, academics, extension agencies, farmers, and other stakeholders to collaborate on their own efforts. Unlocking the full potential of integrated water management and creating a more sustainable future for agriculture and water management around the world can be accomplished through the promotion of awareness, the provision of technical support, the implementation of policies that are supportive, the promotion of collaboration, and resilience to climate change.

References

Ayars, J.E., Hutmacher, R.B., Schoneman, R.A., Soppe, R.W.O., Vail, S.S., & Dale, F. (1999). Realizing the potential of integrated irrigation and drainage water management for meeting crop water requirements in semi-arid and arid areas. Irrigation and Drainage Systems, 13(4), 321-347.

Bouwer, H. (2000). Integrated water management: emerging issues and challenges. Agricultural Water Management, 45(3), 217-228.

Chartzoulakis, K., and Bertaki, M. (2015). Sustainable water management in agriculture under climate change. Agriculture and Agricultural Science Procedia, 4, 88-98.

Falloon, P., and Betts, R. (2010). Climate impacts on European agriculture and water management in the context of adaptation and mitigation—the importance of an integrated approach. Science of the Total Environment, 408(23), 5667-5687.

Jägermeyr, J., Gerten, D., Schaphoff, S., Heinke, J., Lucht, W., and Rockström, J. (2016). Integrated crop water management might sustainably halve the global food gap. Environmental Research Letters, 11(2), 025002.

Qadir, M., Boers, T. M., Schubert, S., Ghafoor, A. and Murtaza, G. (2003). Agricultural water management in water-starved countries: challenges and opportunities. Agricultural Water Management, 62(3), 165-185.

16

Integrated Water Management in Horticultural Crops

Pankaj Kumar[1], Pradipto Kumar Mukherjee[2], Anushi[3] and Durgesh Kumar Maurya[4]

Department of Agriculture Engineer (Process and Food Engineering) SVPUAT, Meerut, Uttar Pradesh
[2]Department Fruit Science, Palli Siksha Bhavana (Institute of Agriculture) Visva –Bharati, West Bengal
[3]Department of Fruit Science, CSAUAT, Kanpur, Uttar Pradesh
[4]Department of Agronomy, CSAUAT, Kanpur, Uttar Pradesh

Abstract

Integrated water management, often known as IWM, is being more acknowledged as an essential component of environmentally responsible crop production in the horticulture industry. Horticultural crops, which include ornamentals, fruits, and vegetables, are very reliant on water in order to achieve their full potential in terms of development and yield. Nevertheless, techniques that are inefficient in terms of water management may result in the waste of water, the deterioration of soil, and contamination of the environment. Integrated Water Management (IWM) provides a comprehensive strategy for addressing these difficulties by using a variety of water management approaches, including irrigation scheduling, water-saving devices, soil moisture monitoring, and conservation measures. The ideas and methods of integrated water management (IWM) in horticultural crops are investigated in this abstract, along with the advantages, problems, and possible solutions associated with IWM. IWM plays a critical role in assuring the long-term sustainability and productivity of horticultural crop production systems. It does this by maximizing the efficiency with which water is used, boosting crop resistance to water stress, and encouraging sustainable water resource management. The implementation of IWM methods has the potential to contribute to a more resilient, productive, and ecologically sustainable horticulture industry. This may be accomplished via cooperation between academics, farmers, policymakers, and other stakeholders.

Keywords: *environment, problems, crop, potential, water*

Introduction

The yearly rainfall in India is 119.4 centimeters, which is equivalent to 391.63 millimeters of rainfall over 328 hectares of land. It is possible to round this up to 400 mham, taking into account the snowfall. Rainwater escapes by evaporation, seeps into the ground, and eventually flows off over the surface of the land. A total of 75 mham of ground water undergoes evaporation on each of the 130 rainy days, while the remaining 55 days suffer evaporation of a comparable magnitude. Seventy millimeters of precipitation are lost to the sky out of the annually average rainfall of four hundred millimeters, with 115 millimeters running off the surface and 215 millimeters soaking into the earth. According to the Irrigation Commission of India (1972), the total annual surface runoff is 180 mham. This figure takes into account 20 mham from stream and river catchments, 45 mham from regenerated ground water, and 115 mham from precipitation. The evaporation rate from medium and big reservoirs is 20%, while the evaporation rate from tanks is 40%. All of this surface water is held in reservoirs and tanks, which total around 15 mham. The remaining 150 mham is distributed to the sea and nations that are adjacent to it. Approximately 215 mham of the yearly rainfall of 400 mham actually makes its way into the soil, while the remaining 165 mham is kept as moisture in the soil. It is estimated that 12.5% of the total precipitation makes its way into the ground water table each year. This is because water that is in excess of the field capacity percolates down and contributes to the ground water. The contribution of rivers and streams to groundwater will be around sixty million cubic meters throughout the course of time as a result of technological advancements. It is estimated that the amount of transpiration received from irrigated and unirrigated crops is 13 mham and 42 mham, respectively, while the amount of transpiration received from forests and other vegetation is 55 mham. After the irrigation system has reached its maximum potential, it is anticipated that the amount of transpiration will be 125 mham, with 35 mham coming from irrigated crops, 35 mham coming from unirrigated crops, and 55 mham coming from forests and other vegetations.

Horticultural crops, encompassing fruits, vegetables, herbs, and ornamental plants, play a vital role in global agriculture, contributing significantly to food security, nutrition, and economic prosperity. However, the sustainable production of horticultural crops faces numerous challenges, including water scarcity, inefficient water use, soil degradation, and environmental pollution. Integrated water management (IWM) offers a comprehensive approach to address these challenges by integrating various water management techniques to optimize water use efficiency, enhance crop productivity, and promote sustainable water resource management in horticultural crop production systems. This article explores the principles, strategies, benefits, challenges, and opportunities of IWM in horticultural crops, drawing on scientific research, case studies, and best practices from around the world.

Water Management Challenges in Horticultural Crop Production

1. **Water Scarcity**: One of the primary challenges in horticultural crop production is water scarcity, particularly in arid and semi-arid regions where water resources are limited. Competition for water among different sectors, including agriculture, industry, and urban areas, exacerbates the problem, leading to inadequate water availability for horticultural crops.
2. **Variability in Rainfall Patterns**: Variability in rainfall patterns, including irregular precipitation, droughts, and erratic weather conditions, poses a significant challenge to water management in horticultural crop production. Inconsistencies in rainfall can lead to water stress in crops, affecting their growth, development, and yield.
3. **Inefficient Irrigation Practices**: Inefficient irrigation practices, such as over-irrigation, under-irrigation, and improper scheduling, contribute to water wastage, soil degradation, and reduced crop productivity. Traditional irrigation methods, such as flood irrigation, often result in water runoff, evaporation, and leaching of nutrients, leading to inefficiencies in water use.
4. **Soil Degradation**: Poor soil management practices, including soil erosion, compaction, and salinization, can exacerbate water management challenges in horticultural crop production. Degraded soils have reduced water-holding capacity, increased runoff, and decreased infiltration rates, leading to water loss and nutrient depletion.

The environmental consequences of water mismanagement in horticultural crop production include:

1. **Soil Erosion**: Inadequate water management practices, such as over-irrigation and improper drainage, can lead to soil erosion, where topsoil is washed away by water runoff. Soil erosion reduces soil fertility, decreases crop productivity, and contributes to land degradation.
2. **Water Pollution**: Poor irrigation practices, such as the excessive use of fertilizers and pesticides, can result in water pollution. Agricultural runoff containing sediment, nutrients, and agrochemicals can contaminate water bodies, leading to eutrophication, algal blooms, and degradation of aquatic ecosystems.

3. **Habitat Loss**: Water mismanagement in horticultural crop production can lead to habitat loss and degradation, particularly in wetland and riparian areas. Excessive water extraction, drainage, and alteration of natural watercourses can disrupt aquatic habitats, endangering biodiversity and ecosystem services.

The economic impacts of water scarcity on horticultural crop production include

1. **Reduced Crop Yields**: Water scarcity can lead to reduced crop yields and quality, affecting farm profitability and income generation for farmers. Water-stressed crops are more susceptible to pests, diseases, and physiological disorders, further reducing yields and marketable produce.
2. **Increased Production Costs**: Farmers may incur additional costs for water acquisition, irrigation infrastructure, and water-saving technologies to cope with water scarcity. Higher production costs reduce farm profitability and competitiveness in the market, particularly for smallholder farmers with limited resources.
3. **Rural Livelihoods**: Water scarcity in horticultural crop production can undermine rural livelihoods, particularly in rural communities dependent on agriculture for income and food security. Reduced crop yields and farm incomes can lead to poverty, unemployment, and migration, exacerbating socio-economic inequalities in rural areas.

Principles of Integrated Water Management

Integrated Water Management (IWM) is a comprehensive approach to water resource management that aims to optimize water use efficiency, minimize water wastage, and promote sustainable water resource management in horticultural crop production systems. The core principles of IWM include

1. **Holistic Approach**: IWM considers the entire water cycle, from water sourcing to irrigation, drainage, and water reuse or recycling. It takes into account the interactions between water, soil, crops, climate, and other environmental factors to develop integrated solutions that address multiple aspects of water management.
2. **Multiple Water Management Techniques**: IWM integrates a variety of water management techniques, including irrigation scheduling, water-saving technologies, soil moisture monitoring, conservation practices,

rainwater harvesting, and crop selection. By combining different techniques, IWM seeks to optimize water use efficiency and minimize water losses throughout the crop production cycle.

3. **Site-Specific Strategies**: IWM recognizes that water management needs vary depending on factors such as soil type, climate, crop type, and local water availability. Site-specific strategies are tailored to the unique characteristics of each agricultural site, optimizing water use and minimizing environmental impacts.
4. **Adaptive Management**: IWM emphasizes adaptive management approaches that allow for flexibility and adjustment in response to changing conditions, such as fluctuating water availability, climate variability, and evolving crop needs. Continuous monitoring, evaluation, and adjustment of water management practices are essential for optimizing performance and resilience over time.

IWM aims to optimize water use efficiency, minimize water wastage, and promote sustainable water resource management in horticultural crop production systems by

1. **Optimizing Water Use Efficiency**: IWM seeks to maximize the productivity of water by ensuring that crops receive the right amount of water at the right time. Techniques such as drip irrigation, precision irrigation, and soil moisture monitoring help deliver water directly to the root zone of plants, minimizing evaporation and runoff and maximizing water uptake efficiency.
2. **Minimizing Water Wastage**: IWM aims to reduce water losses through evaporation, runoff, and deep percolation. Practices such as mulching, cover cropping, and conservation tillage help conserve soil moisture, reduce water runoff, and minimize water wastage, especially in areas with limited water availability.
3. **Promoting Sustainable Water Resource Management**: IWM promotes sustainable water resource management by integrating water conservation practices, promoting water reuse and recycling, and protecting water quality. Techniques such as rainwater harvesting, irrigation system upgrades, and soil health management contribute to the sustainable use and conservation of water resources.

Examples of successful IWM initiatives and projects in horticultural crop production include

- Implementation of drip irrigation systems in fruit orchards to improve water use efficiency and crop yields.
- Adoption of soil moisture sensors and automated irrigation controllers to optimize irrigation scheduling and reduce water wastage in vegetable production.
- Integration of cover cropping and conservation tillage practices to enhance soil moisture retention and reduce erosion in vineyards.
- Establishment of rainwater harvesting systems in greenhouse operations to supplement irrigation water sources and reduce reliance on freshwater supplies.

Components of Integrated Water Management in Horticultural Crops

Integrated Water Management (IWM) in horticultural crops encompasses a range of components aimed at optimizing water use efficiency, minimizing water wastage, and promoting sustainable water resource management. Here are the key components of IWM in horticultural crops:

1. Irrigation Scheduling and Management

- Irrigation scheduling involves determining when and how much water to apply to crops based on factors such as crop water requirements, soil moisture levels, weather conditions, and stage of crop growth.
- Proper irrigation management ensures that crops receive adequate water without excess, minimizing water wastage and reducing the risk of water stress or waterlogging.
- Techniques for irrigation scheduling and management include using weather-based models, soil moisture sensors, and crop coefficients to optimize irrigation timing and frequency.

2. Water-saving Technologies

- Water-saving technologies aim to improve irrigation efficiency and reduce water losses through evaporation, runoff, and deep percolation.
- Examples of water-saving technologies include drip irrigation, micro-sprinklers, and sub-surface irrigation systems, which deliver water directly to the root zone of plants, minimizing losses from evaporation and runoff.
- Other technologies, such as laser leveling, precision land grading, and surge valves, help optimize water distribution and minimize water losses in irrigation systems.

3. Soil Moisture Monitoring

- Soil moisture monitoring involves measuring and monitoring the moisture content of the soil to assess water availability for crops and guide irrigation decisions.
- Soil moisture sensors, tensiometers, and neutron probes are commonly used tools for soil moisture monitoring, providing real-time data on soil moisture levels and helping farmers optimize irrigation scheduling.
- Monitoring soil moisture helps prevent both under-irrigation, which can lead to water stress in crops, and over-irrigation, which can waste water and cause leaching of nutrients.

4. Conservation Practices

- Conservation practices aim to conserve soil moisture, reduce water runoff, and minimize soil erosion, thus improving water use efficiency in horticultural crop production.
- Techniques such as mulching, cover cropping, contour farming, and conservation tillage help retain soil moisture, improve soil structure, and reduce water runoff and erosion.
- Conservation practices also contribute to soil health and fertility, enhancing the overall sustainability of horticultural crop production systems.

5. Rainwater Harvesting

- Rainwater harvesting involves capturing and storing rainwater for irrigation purposes, supplementing conventional water sources and reducing reliance on groundwater or surface water.
- Techniques for rainwater harvesting include rooftop rainwater collection, surface runoff collection, and harvesting runoff from agricultural fields.
- Rainwater harvesting systems can range from simple rain barrels to more complex storage tanks and reservoirs, depending on the scale of operations and water requirements.

6. Crop Selection

- Crop selection plays a crucial role in IWM, as different crops have varying water requirements, drought tolerance, and suitability to local growing conditions.

- Selecting drought-tolerant or water-efficient crop varieties can help minimize water use and reduce the risk of water stress in horticultural crop production.
- Crop rotation and diversification strategies also contribute to IWM by balancing water demand, improving soil health, and reducing pest and disease pressures.

Examples of innovative technologies and practices used in IWM for horticultural crops include

- Drip irrigation systems, which deliver water directly to the root zone of plants, minimizing water wastage and optimizing water use efficiency.
- Soil moisture sensors and automated irrigation controllers, which provide real-time data on soil moisture levels and adjust irrigation schedules accordingly to prevent water stress and minimize water wastage.
- Cover cropping and conservation tillage practices, which help retain soil moisture, improve soil structure, and reduce water runoff and erosion, thus enhancing water use efficiency and soil health

Benefits of Integrated Water Management in Horticultural Crops

1. Improved Water Use Efficiency

- IWM optimizes water use efficiency by ensuring that crops receive the right amount of water at the right time, minimizing water losses through evaporation, runoff, and deep percolation.
- Studies have shown that implementing IWM practices such as drip irrigation, soil moisture monitoring, and conservation practices can improve water use efficiency by up to 50% or more compared to conventional irrigation methods (FAO, 2018).
- Real-world examples demonstrate significant water savings and increased crop yields through the adoption of IWM practices in horticultural crop production (Bhattarai et al., 2019).

2. Enhanced Crop Productivity and Quality

- IWM practices contribute to enhanced crop productivity and quality by providing crops with optimal moisture levels and reducing water stress during critical growth stages.

- Research has shown that properly managed irrigation using IWM techniques can increase crop yields by 20-50% or more while improving fruit size, color, flavor, and nutritional content (Schwarz et al., 2020).
- Case studies from horticultural crop production systems worldwide demonstrate the positive impacts of IWM on crop yields, quality, and marketability (Abd-Elgawad et al., 2021).

3. Reduced Water Wastage and Pollution

- IWM helps minimize water wastage and pollution by reducing runoff, leaching of nutrients and agrochemicals, and contamination of water bodies.
- Studies have shown that implementing IWM practices such as drip irrigation, mulching, and cover cropping can significantly reduce water runoff, soil erosion, and nutrient leaching, leading to improved water quality and ecosystem health (Savio et al., 2018).
- Real-world examples demonstrate the effectiveness of IWM in reducing water pollution and improving water resource management in horticultural crop production systems (Gan et al., 2020).

4. Environmental Sustainability

- IWM contributes to environmental sustainability by conserving water resources, preserving soil health, and minimizing the negative impacts of agriculture on ecosystems and biodiversity.
- Adoption of IWM practices such as rainwater harvesting, conservation tillage, and crop rotation promotes soil conservation, biodiversity, and ecosystem services, enhancing the resilience of agricultural landscapes to climate change and other environmental challenges (FAO, 2019).
- Case studies from sustainable agriculture initiatives around the world highlight the role of IWM in promoting environmental sustainability and resilience in horticultural crop production (Guzmán et al., 2021).

5. Economic Viability

- IWM practices contribute to economic viability and profitability by reducing input costs, improving crop yields and quality, and enhancing market competitiveness.

- Studies have shown that investing in IWM technologies and practices can generate positive returns on investment, with significant cost savings from reduced water usage, labour, and agrochemical inputs (FAO, 2020).
- Real-world examples demonstrate the economic benefits of IWM for farmers, agribusinesses, and rural communities, including increased farm income, job creation, and market access (Kisekka et al., 2021).

Challenges and Barriers to Integrated Water Management

1. **Limited Awareness and Education**: Many farmers lack awareness and understanding of IWM principles, techniques, and benefits. Limited access to information and extension services further exacerbates this challenge, leading to low adoption rates of IWM practices.
2. **High Initial Investment Costs**: Implementing IWM practices often requires significant upfront investment in infrastructure, technologies, and training. High initial costs can deter farmers, especially smallholders with limited financial resources, from adopting IWM practices.
3. **Technical Complexity**: IWM practices may involve technical complexities, such as the installation and maintenance of irrigation systems, soil moisture monitoring equipment, and water-saving technologies. Farmers may lack the technical knowledge and skills required to implement and manage these practices effectively.
4. **Access to Resources**: Limited access to water resources, land, capital, and inputs such as irrigation equipment, soil moisture sensors, and drought-resistant crop varieties can hinder the adoption of IWM practices, particularly among resource-constrained farmers.
5. **Policy and Institutional Barriers**: Inadequate policy support, unclear regulations, and bureaucratic barriers can impede the adoption of IWM practices. Lack of coordination among government agencies, conflicting policies, and institutional inertia may hinder efforts to promote IWM at the national and local levels.
6. **Social and Cultural Factors**: Socio-cultural factors, including traditional farming practices, gender norms, and farmer perceptions, can influence the adoption of IWM practices. Resistance to change, cultural preferences, and social norms may hinder the uptake of new technologies and approaches.
7. **Water Governance Issues**: Weak water governance frameworks, lack of water rights, and competition for water resources among different

stakeholders can pose challenges to implementing IWM practices. Inefficient water allocation, over-extraction, and water conflicts may further exacerbate water management challenges.

8. **Climate Change Impacts**: Climate change exacerbates water management challenges in horticultural crop production, leading to increased water variability, extreme weather events, and unpredictable rainfall patterns. Climate-related risks, such as droughts, floods, and heatwaves, can affect crop yields, water availability, and livelihoods.

Conclusion

In conclusion, Integrated Water Management (IWM) holds immense potential for addressing water management challenges and promoting sustainable horticultural crop production. Despite facing various challenges and barriers, the adoption of IWM practices offers numerous benefits, including improved water use efficiency, enhanced crop productivity and quality, reduced water wastage and pollution, environmental sustainability, and economic viability. By embracing a holistic approach that integrates multiple water management techniques, such as irrigation scheduling and management, water-saving technologies, soil moisture monitoring, conservation practices, rainwater harvesting, and crop selection, stakeholders can optimize water use, mitigate environmental impacts, and enhance resilience to climate change. However, overcoming challenges such as limited awareness and education, high initial investment costs, technical complexity, access to resources, policy and institutional barriers, social and cultural factors, water governance issues, and climate change impacts requires concerted efforts and collaboration among governments, farmers, researchers, NGOs, and other stakeholders. By implementing strategies such as increasing awareness and education, providing financial support and incentives, promoting policy reforms, and fostering multi-stakeholder collaboration, stakeholders can unlock the full potential of IWM in horticultural crop production. Ultimately, by embracing IWM principles and practices, we can promote sustainable water management, ensure food security, enhance rural livelihoods, and safeguard the environment for present and future generations.

References

Bouwer, H. (2000). Integrated water management: emerging issues and challenges. Agricultural water management, 45(3), 217-228.

Garcia-Caparros, P., Contreras, J. I., Baeza, R., Segura, M. L., & Lao, M. T. (2017). Integral management of irrigation water in intensive horticultural systems of Almería. Sustainability, 9(12), 2271.

Incrocci, L., Thompson, R. B., Fernandez-Fernandez, M. D., De Pascale, S., Pardossi, A., Stanghellini, C., .& Gallardo, M. (2020). Irrigation management of European greenhouse vegetable crops. Agricultural Water Management, 242, 106393.

Karuna, K., & Mankar, A. (2020). Water Management in Horticultural Crops. In Sustainable Agriculture (pp. 211-226). Apple Academic Press.

Zinkernagel, J., Maestre-Valero, J. F., Seresti, S. Y., & Intrigliolo, D. S. (2020). New technologies and practical approaches to improve irrigation management of open field vegetable crops. Agricultural Water Management, 242, 106404.

17

Agricultural Extension Services for Better Management of Produce

P. Kalaiselvi[1], Pragati Prasad Dessai[2], Anshika Jain[3] and Nisha Yadav[4]

[1](ENS), ICAR- Krishi Vigyan Kendra, Tamil Nadu Agricultural University Salem, Tamil Nadu
[2]Agriculture Extension Education and Communication, Sam Higginbotham University of Agriculture Sciences and Technology, Prayagraj, Uttar Pradesh
[3]Agriculture Extension Education and Communication, Sam Higginbotham University of Agriculture Sciences and Technology, Prayagraj, Uttar Pradesh
[4]Agricultural Extension Education, SVPUAT, Meerut, Uttar Pradesh

Abstract

Agricultural extension services play a pivotal role in facilitating better management of produce by providing farmers with knowledge, resources, and support to improve their agricultural practices. These services act as a bridge between agricultural research and farmers, disseminating information on innovative techniques, technologies, and best practices tailored to local conditions. Through workshops, training sessions, field demonstrations, and one-on-one consultations, extension agents empower farmers to adopt sustainable production methods, optimize inputs, and enhance crop yields while minimizing environmental impact. By promoting efficient post-harvest handling, storage, and marketing strategies, extension services help farmers reduce post-harvest losses, improve product quality, and access lucrative markets. Additionally, extension programs often incorporate valuable information on crop rotation, integrated pest management, soil conservation, and water management practices, enabling farmers to make informed decisions for long-term agricultural sustainability. Ultimately, agricultural extension services serve as catalysts for innovation, productivity, and resilience in the agricultural sector, contributing to food security, economic development, and rural livelihoods.

Keywords: *Development, services, extension, management*

Introduction

Agricultural extension services serve as a vital component of agricultural development by bridging the gap between research institutions and farmers. These services play a crucial role in disseminating knowledge, providing technical assistance, and facilitating capacity-building initiatives to improve agricultural practices and enhance productivity. In the context of produce management, agricultural extension services offer a wealth of resources and support aimed at optimizing post-harvest processes, reducing losses, and maximizing the value of agricultural produce. This article explores the multifaceted role of agricultural extension services in promoting better management of produce, highlighting their significance in achieving sustainable agricultural outcomes.

Historical Evolution of Agricultural Extension Services

The concept of agricultural extension services traces its roots back to the late 19th and early 20th centuries when agricultural societies, government agencies, and educational institutions began to recognize the need for disseminating scientific knowledge and best practices to farmers. Early extension efforts focused primarily on improving crop production techniques, addressing pest and disease management challenges, and promoting the adoption of new technologies such as mechanization and improved seed varieties.

Over time, the scope of agricultural extension services expanded to encompass a broader range of agricultural issues, including soil conservation, water management, livestock husbandry, and post-harvest handling. Extension services evolved from traditional top-down approaches to more participatory and farmer-cantered models, emphasizing collaboration, empowerment, and community engagement. Today, agricultural extension services operate within diverse institutional frameworks, including government agencies, non-governmental organizations (NGOs), research institutions, and private sector entities, each contributing to the delivery of extension programs tailored to the needs of farmers and agricultural communities.

Role of Agricultural Extension Services in Produce Management

Agricultural extension services play a critical role in produce management by bridging the gap between scientific research and practical application on farms. These services support farmers in adopting improved agricultural practices, optimizing post-harvest processes, and enhancing the overall quality and marketability of their produce. Here are the key roles agricultural extension services play in produce management:

1. Dissemination of Knowledge and Best Practices

Educational Outreach: Extension agents educate farmers on the latest research findings and best practices in produce management. This includes information on optimal harvesting techniques, post-harvest handling, and storage methods that minimize losses and maintain product quality.

Field Demonstrations and Workshops: Hands-on training sessions and field demonstrations allow farmers to see best practices in action. These practical learning experiences help farmers understand and implement new techniques effectively.

2. Technical Assistance and Support

Problem-Solving: Extension agents provide technical assistance to help farmers address specific challenges in produce management, such as pest infestations, diseases, or nutrient deficiencies. This support can include diagnostic services, personalized advice, and on-site visits.

Adoption of Technologies: Extension services facilitate the adoption of appropriate technologies that enhance produce management. This can include the use of improved storage facilities, refrigeration units, and packaging materials that extend the shelf life and quality of produce.

3. Post-Harvest Loss Reduction

Training on Post-Harvest Handling: Farmers are trained in proper handling techniques to reduce physical damage to produce during harvesting, sorting, and packaging. This includes careful handling of fruits and vegetables to prevent bruising and spoilage.

Improved Storage Solutions: Extension services promote the use of effective storage solutions, such as cold storage, controlled atmosphere storage, and traditional methods adapted to local conditions. These methods help maintain the freshness and nutritional value of produce.

Transportation Best Practices: Proper transportation methods are critical to reducing losses. Extension agents advise on best practices for transporting produce to markets, including appropriate packaging and temperature control.

4. Market Access and Value Addition

Market Information and Intelligence: Extension services provide farmers with valuable market information, including price trends, demand forecasts, and consumer preferences. This information helps farmers plan their production and marketing strategies more effectively.

Value Addition Techniques: Farmers are trained in value addition techniques, such as processing, packaging, and branding, to increase the market value of their produce. This can include drying fruits, making preserves, or creating branded products that appeal to niche markets.

Linking Farmers to Markets: Extension services facilitate connections between farmers and markets, including local, regional, and international buyers. This can involve organizing farmer markets, establishing cooperatives, and connecting farmers with agribusinesses and retailers.

5. Promotion of Sustainable Practices

Sustainable Farming Practices: Extension services promote sustainable farming practices that improve soil health, conserve water, and enhance biodiversity. Techniques such as crop rotation, organic farming, and integrated pest management are encouraged to create resilient agricultural systems.

Environmental Stewardship: By advocating for environmentally friendly practices, extension services help farmers reduce their environmental footprint. This includes promoting the use of renewable energy sources, such as solar-powered irrigation systems, and minimizing the use of chemical inputs.

6. Capacity Building and Empowerment

Farmer Field Schools: Extension services often establish farmer field schools where farmers can learn from each other and share experiences. These schools provide a platform for collaborative learning and problem-solving.

Empowering Women and Youth: Extension programs specifically target women and young farmers, providing them with the skills and knowledge needed to succeed in agriculture. This empowerment helps diversify income sources and improves the economic stability of rural communities.

7. Policy Advocacy and Support

Advocacy for Farmer-Friendly Policies: Extension services advocate for policies that support farmers, such as subsidies for sustainable inputs, access to credit, and infrastructure development. By influencing policy, extension services help create an enabling environment for successful produce management.

Implementation of Standards and Certification: Extension agents assist farmers in meeting quality standards and obtaining certifications such as organic, fair trade, or GAP (Good Agricultural Practices). These certifications can open up new markets and increase the profitability of produce.

Case Studies Highlighting Successful Produce Management

1. **Kenya: Improving Mango Value Chains:** In Kenya, agricultural extension services have significantly improved the mango value chain. Extension agents introduced farmers to better harvesting techniques, proper sorting, and grading practices, and the use of hot water treatment to reduce post-harvest losses caused by fruit flies. Additionally, farmers were linked to local and international markets, enhancing their income through better prices and reduced wastage.

2. **India: Enhancing Tomato Storage:** In India, the introduction of zero-energy cool chambers through extension services has revolutionized tomato storage. These chambers, which use evaporative cooling, allow farmers to store tomatoes for longer periods without refrigeration, reducing spoilage and extending the marketing window. This innovation has been widely adopted, significantly reducing post-harvest losses and increasing farmers' profits.

3. **Peru: Organic Quinoa Production:** In Peru, extension services have supported the transition to organic quinoa production, a crop native to the Andean region. Farmers received training in organic farming techniques, pest and disease management, and organic certification processes. The result has been an increase in organic quinoa production, higher prices for certified organic produce, and improved livelihoods for smallholder farmers.

Case Studies and Success Stories

Case Studies and Success Stories in Agricultural Extension Services for Better Produce Management

Agricultural extension services have demonstrated significant success worldwide in improving produce management, enhancing farm productivity, and promoting sustainable agricultural practices. Here are detailed case studies and success stories highlighting the impactful role of these services in different regions and contexts.

1. Kenya: Enhancing Mango Value Chains

Background: In Kenya, the mango industry faces challenges such as post-harvest losses due to poor handling and pest infestations, particularly by fruit flies. These issues have historically led to substantial economic losses for farmers.

Intervention: The Kenya Agricultural and Livestock Research Organization (KALRO), in collaboration with the USAID-funded Kenya Agricultural Value Chain Enterprises (KAVES) project, implemented a series of training programs for mango farmers. These programs focused on integrated pest management (IPM) techniques, including the use of bait traps and biological controls to manage fruit fly populations. Additionally, farmers were trained in proper harvesting, sorting, grading, and the use of hot water treatment to reduce post-harvest losses.

Outcome: The intervention led to a significant reduction in fruit fly infestations and post-harvest losses. Farmers who adopted the recommended practices saw an increase in the quality and marketability of their mangoes, leading to higher prices. Moreover, the establishment of linkages with local and international markets helped farmers secure better returns. The success of this initiative has been widely recognized, resulting in its expansion to other mango-growing regions in Kenya.

2. India: Enhancing Tomato Storage with Zero-Energy Cool Chambers

Background: In India, tomato farmers often struggle with post-harvest losses due to inadequate storage facilities and high temperatures, which accelerate spoilage.

Intervention: The Indian Agricultural Research Institute (IARI) introduced zero-energy cool chambers, a low-cost, eco-friendly storage solution that utilizes evaporative cooling to maintain lower temperatures. Extension services played a crucial role in educating farmers about the construction and use of these cool chambers.

Outcome: The adoption of zero-energy cool chambers enabled farmers to store tomatoes for longer periods, reducing spoilage and extending the marketing window. This innovation resulted in a decrease in post-harvest losses by up to 70%, allowing farmers to sell their produce at higher prices. The success of this technology has spurred interest in other perishable crops, enhancing overall agricultural productivity and profitability in the region.

3. Peru: Transitioning to Organic Quinoa Production

Background: Quinoa, a traditional Andean crop, has seen growing international demand due to its nutritional benefits. However, smallholder farmers in Peru faced challenges in meeting organic certification standards and accessing lucrative markets.

Intervention: The International Potato Centre (CIP) and local extension services collaborated to support farmers in transitioning to organic quinoa production. This included training on organic farming practices, pest and disease management, soil fertility enhancement, and the organic certification process. Farmers were also educated on value addition techniques and market linkages.

Outcome: The transition to organic production increased the market value of quinoa, providing higher income for farmers. Organic certification opened up new market opportunities, both locally and internationally. The success of this initiative has empowered smallholder farmers, improved livelihoods, and promoted sustainable farming practices in the Andean region.

4. Bangladesh: Empowering Women through Integrated Pest Management

Background: In Bangladesh, rice farmers face significant yield losses due to pest infestations. Women, who play a crucial role in rice farming, often lack access to knowledge and resources for effective pest management.

Intervention: The Bangladesh Rice Research Institute (BRRI), in partnership with local extension services, implemented a program focused on integrated pest management (IPM). The program included training sessions specifically targeted at women farmers, teaching them about biological pest control methods, the use of pheromone traps, and cultural practices that reduce pest incidence.

Outcome: The program successfully empowered women farmers with the knowledge and skills needed to implement IPM practices. As a result, rice yields increased, and pesticide use was significantly reduced, leading to cost savings and improved health outcomes. The program not only enhanced agricultural productivity but also promoted gender equality by involving women in decision-making processes related to farm management.

5. Nigeria: Improving Cassava Post-Harvest Processing

Background: Cassava is a staple crop in Nigeria, but post-harvest losses due to inadequate processing and storage have been a major challenge.

Intervention: The International Institute of Tropical Agriculture (IITA) and local extension services introduced improved post-harvest processing techniques and storage facilities to cassava farmers. Training sessions focused on processing cassava into higher-value products like gari, fufu, and cassava flour. Additionally, farmers were educated on the use of improved storage methods to extend the shelf life of cassava products.

Outcome: The introduction of improved processing techniques and storage facilities significantly reduced post-harvest losses and increased the income of cassava farmers. By adding value to their produce, farmers were able to access new markets and secure better prices. This initiative also enhanced food security by ensuring a more stable supply of cassava products.

Challenges

Agricultural extension services play a crucial role in enhancing produce management by providing farmers with the necessary knowledge, technical assistance, and support. However, these services face numerous challenges that can hinder their effectiveness. Understanding these challenges is essential for improving the delivery and impact of extension services. Key challenges include:

1. Resource Constraints

Limited Funding: Extension services often operate with insufficient funding, affecting their ability to reach farmers, particularly in remote and underserved areas. Budget constraints can limit the availability of resources such as training materials, transportation, and extension agents.

Inadequate Staffing: Many extension programs suffer from a shortage of trained personnel. The ratio of extension agents to farmers is often too low, making it difficult to provide personalized and timely support.

Infrastructure Deficiencies: Lack of adequate infrastructure, such as offices, demonstration farms, and transportation, hampers the ability of extension agents to conduct effective outreach and training activities.

2. Capacity and Training of Extension Agents

Skills and Knowledge Gaps: Extension agents may lack up-to-date knowledge and skills in modern agricultural practices and technologies. Continuous professional development and training are often insufficient, limiting their ability to deliver high-quality advice.

Motivation and Incentives: Extension agents might be poorly compensated and lack motivation due to inadequate incentives and career advancement opportunities. This can lead to low job satisfaction and high turnover rates.

3. Information Dissemination and Communication

Reaching Remote Areas: Accessing farmers in remote and geographically challenging areas is a significant barrier. Poor infrastructure and logistical challenges make it difficult for extension agents to provide regular and effective support.

Use of Technology: While modern communication technologies can enhance information dissemination, many farmers and extension agents lack access to these technologies or the skills to use them effectively.

Language and Cultural Barriers: Communication can be hindered by language differences and cultural barriers. Extension agents need to tailor their communication strategies to the local context and ensure that materials are accessible and understandable to all farmers.

4. Farmer Engagement and Participation

Trust and Relationship Building: Building trust and strong relationships with farmers is essential but can be challenging. Farmers may be skeptical of new practices or wary of external advice, particularly if past experiences have been negative.

Farmer Heterogeneity: Farmers have diverse needs, preferences, and circumstances. Extension services must be flexible and adaptive to cater to different types of farmers, including smallholders, women, and youth.

Participation and Empowerment: Encouraging active participation and empowerment of farmers in the learning process is crucial for effective knowledge transfer. Extension services must foster an inclusive environment where farmers feel valued and heard.

5. Integration and Coordination

Fragmentation of Services: Extension services are often fragmented and lack coordination among various stakeholders, including government agencies, NGOs, and private sector actors. This can lead to duplication of efforts and inefficient use of resources.

Policy and Institutional Support: Weak policy frameworks and institutional support can limit the effectiveness of extension services. Stronger policies and better institutional coordination are needed to create an enabling environment for extension activities.

Sustainability and Continuity: Ensuring the sustainability and continuity of extension programs is challenging. Short-term projects and funding cycles can disrupt ongoing efforts and diminish long-term impacts.

6. Market and Economic Challenges

Market Linkages: Extension services often struggle to create effective market linkages for farmers. Helping farmers access markets and achieve better prices for their produce requires strong connections with value chain actors and market information systems.

Economic Viability: Advising farmers on practices that are economically viable and sustainable is challenging. Extension agents must balance the promotion of innovative practices with the economic realities faced by farmers.

7. Evaluation and Impact Assessment

Monitoring and Evaluation: Measuring the impact of extension services on produce management and overall agricultural productivity is complex. Robust monitoring and evaluation systems are needed to assess effectiveness and guide improvements.

Data Collection and Analysis: Collecting and analysing data to inform extension activities and policies requires resources and technical expertise that may be lacking.

Strategies to Overcome Challenges

Increased Investment: Governments and development partners should increase funding for extension services to improve infrastructure, staffing, and resource availability.

Capacity Building: Continuous training and professional development for extension agents are essential to keep them updated on the latest agricultural practices and technologies.

Use of ICT: Leveraging information and communication technologies (ICT) can enhance outreach and information dissemination, especially in remote areas.

Community Engagement: Building trust and fostering strong relationships with farmers through participatory approaches can improve the effectiveness of extension services.

Integrated Approaches: Coordination among various stakeholders, including government, NGOs, and the private sector, can create synergies and improve the overall impact of extension services.

Policy Support: Strengthening policy frameworks and institutional support can create an enabling environment for effective extension services.

Market-Oriented Extension: Developing market linkages and providing market information can help farmers make informed decisions and improve their economic outcomes.

Robust M&E Systems: Implementing strong monitoring and evaluation systems can help assess the impact of extension services and guide continuous improvements.

Conclusion

Agricultural extension services play a pivotal role in promoting better management of produce by providing farmers with knowledge, resources, and support to optimize post-harvest processes, reduce losses, and maximize the value of agricultural produce. Through dissemination of best practices, technical assistance, post-harvest loss reduction initiatives, market access facilitation, and promotion of sustainable practices, extension services contribute to enhanced agricultural productivity, improved food security, and sustainable rural development. By harnessing the power of agricultural extension services, stakeholders can work together to build resilient and inclusive agricultural systems that benefit farmers, consumers, and the environment alike.

References

Al-Zahrani, K. H., Aldosari, F. O., Baig, M. B., Shalaby, M. Y., & Straquadine, G. S. (2016). Role of Agricultural Extension Service in Creating Decisionmaking Environment for the Farmers to Realize Sustainable Agriculture in Al-Qassim and Al-Kharj Regions-Saudi Arabia. Japs: Journal of Animal & Plant Sciences, 26(4).

Ali, A. S., Altarawneh, M., & Altahat, E. (2012). Effectiveness of agricultural extension activities. American Journal of Agricultural and Biological Sciences, 7(2), 194-200.

Antwi-Agyei, P., & Stringer, L. C. (2021). Improving the effectiveness of agricultural extension services in supporting farmers to adapt to climate change: Insights from northeastern Ghana. Climate Risk Management, 32, 100304.

Debelo, W. B., Minale, M., Fisha, H., Tedla, A., Eshetu, R., Jafar, M., & Wakgari, G. (2020). Review on roles and challenges of agricultural extension system on growth of agricultural production in Ethiopia. Journal of Plant Sciences, 8(6), 189-200.

Kaur, K., & Kaur, P. (2018). Agricultural extension approaches to enhance the knowledge of farmers. International Journal of Current Microbiology and Applied Sciences, 7(2), 2367-2376.

Laurent, C., Cerf, M., & Labarthe, P. (2006). Agricultural extension services and market regulation: learning from a comparison of six EU countries. Journal of Agricultural Education and Extension, 12(1), 5-16.

Mahaliyanaarachchi, R. P., & Bandara, R. M. A. S. (2006). Commercialization of agriculture and role of agricultural extension. Sabaragamuwa University Journal, 6(1), 13-22.

18

Advanced Marketing of Agricultural Produce

Abhishek Singh[1], Kushal Sachan[2], Vikash Mishra[3] and Susmi Biswas[4]

[1]Department of Agricultural Economics, Chandra Shekhar Azad University of Agriculture and Technology, Kanpur, Uttar Pradesh
[2]Department of Soil Science and Agricultural Chemistry, C. S. Azad University of Agriculture and Technology, Kanpur, Uttar Pradesh
[3]Department of Agricultural Economics, Maharshi Uttam Samajhik Vigyan Kendra, Sehore, Madhya Pradesh
[4]Haldia Institute of Management, Maulana Abul Kalam Azad University of Technology, West Bengal

Abstract

Advanced marketing of agricultural produce encompasses innovative strategies and technologies aimed at enhancing the efficiency, reach, and profitability of agricultural markets. This approach integrates modern tools such as digital platforms, precision marketing, and data analytics to better connect farmers with consumers and markets. It also emphasizes the importance of value addition, branding, and sustainable practices to meet evolving consumer preferences and market demands. By leveraging these advanced marketing techniques, agricultural producers can optimize their supply chains, reduce post-harvest losses, and achieve higher economic returns, ultimately contributing to a more resilient and sustainable agricultural sector.

Keywords: *encompasses, data, consumers, markets, economics*

Introduction

The agricultural sector has long been the backbone of many economies, especially in developing countries where it contributes significantly to GDP and employment. Traditionally, the marketing of agricultural produce has been rudimentary, relying on local markets and middlemen to connect farmers with consumers. However, with globalization, technological advancements, and changing consumer preferences, there is an increasing need to modernize and advance the marketing strategies for agricultural produce. Advanced marketing integrates modern tools, technologies, and innovative practices to enhance the efficiency, reach, and profitability of agricultural markets. This comprehensive article explores the various dimensions of advanced marketing in agriculture, the benefits it offers, the challenges it poses, and successful case studies from around the world.

Challenges in Traditional Agricultural Marketing

Traditional agricultural marketing faces numerous challenges that hinder the efficiency, profitability, and sustainability of the agricultural sector. These challenges impact both small-scale and large-scale farmers, leading to reduced income, increased post-harvest losses, and limited market access. Understanding these challenges is crucial for developing strategies to improve agricultural marketing systems. The key challenges in traditional agricultural marketing include:

1. Fragmented Markets

Small and Disconnected Farms: Agricultural markets are often highly fragmented, with numerous small-scale farmers operating independently. This fragmentation leads to inefficiencies, such as duplication of efforts and an inability to achieve economies of scale. Small farms typically produce lower quantities of produce, making it difficult to meet the demand of larger markets or secure better pricing.

Lack of Coordination: Fragmented markets lack coordination among producers, which can lead to inconsistent supply, varying quality of produce, and difficulties in collective bargaining. This disorganization prevents farmers from benefiting from collective marketing strategies that could reduce costs and increase bargaining power.

2. Dependency on Middlemen

Price Exploitation: Farmers often rely on intermediaries or middlemen to sell their produce. These middlemen can exploit their position, offering low prices to farmers while taking a significant margin for themselves. This dependency reduces the farmers' share of the final sale price and diminishes their income.

Limited Negotiation Power: Without direct access to markets, farmers have limited negotiation power. They often have to accept the prices offered by middlemen, even if they are unfavorable. This can be particularly challenging during harvest seasons when the supply is high, and the urgency to sell perishable goods increases.

3. Post-Harvest Losses

Inadequate Storage: Poor storage facilities and techniques lead to significant post-harvest losses. Perishable goods like fruits and vegetables are especially vulnerable to spoilage if not stored properly. These losses reduce the overall quantity of produce available for sale and thus the potential income for farmers.

Suboptimal Handling and Transportation: Inadequate handling and transportation infrastructure exacerbate post-harvest losses. Poor roads, lack of refrigeration, and inefficient logistics mean that produce often spoils before reaching the market. This not only impacts the farmers' income but also affects food security.

4. Market Information Asymmetry

Lack of Price Information: Farmers often lack timely and accurate information about market prices, demand, and supply trends. This information asymmetry leads to suboptimal decision-making, such as planting crops that are not in high demand or selling produce at lower prices than could be obtained.

Difficulty in Predicting Market Trends: Without access to reliable market data, farmers struggle to predict market trends and make informed decisions about what to plant and when to sell. This unpredictability can lead to overproduction or underproduction, both of which negatively impact income.

5. Limited Market Access

Geographical Barriers: Many small-scale farmers, especially those in remote areas, face significant geographical barriers that limit their access to larger, more lucrative markets. Poor infrastructure, such as roads and transportation networks, further restricts their ability to reach buyers.

Market Exclusivity: Larger markets and retailers often require large quantities of consistent quality produce, which small-scale farmers struggle to meet. This exclusivity restricts access to higher-paying markets and forces farmers to sell in local markets at lower prices.

6. Lack of Value Addition and Branding

Minimal Processing: In traditional marketing systems, there is often little value addition to the raw agricultural produce. Minimal processing or packaging means that farmers miss out on potential additional revenue streams that come from selling higher-value products.

Weak Branding: Farmers typically do not have the resources or knowledge to develop strong brands for their produce. Without branding, it is difficult to differentiate their products in the market, leading to lower consumer recognition and loyalty.

7. Regulatory and Policy Constraints

Inconsistent Policies: Regulatory and policy constraints, such as inconsistent government policies and market regulations, can create uncertainties for farmers. Frequent changes in export policies, price controls, and subsidies disrupt market stability and planning.

Lack of Supportive Infrastructure: Insufficient investment in supportive infrastructure such as storage facilities, transportation networks, and market facilities hampers the effective marketing of agricultural produce. This lack of infrastructure leads to inefficiencies and increased costs.

8. Financial Constraints

Limited Access to Credit: Small-scale farmers often face difficulties in accessing credit and financial services. Without adequate financing, they cannot invest in improved agricultural practices, storage facilities, or marketing strategies that could enhance their market position.

High Cost of Inputs: The high cost of agricultural inputs such as seeds, fertilizers, and pesticides, combined with low selling prices, reduces the profitability of farming. Farmers struggle to break even or make a profit, limiting their ability to invest in better marketing practices.

9. Environmental and Climatic Challenges

Climate Variability: Traditional agricultural marketing is vulnerable to environmental and climatic challenges. Climate variability and extreme weather events can affect crop yields and quality, leading to inconsistent supply and fluctuating prices.

Natural Disasters: Natural disasters such as floods, droughts, and storms can devastate crops and disrupt marketing channels. Recovery from such events is often slow and costly, impacting the farmers' ability to consistently supply markets.

Strategies to Address Challenges in Traditional Agricultural Marketing

To overcome the challenges in traditional agricultural marketing, a multifaceted approach is necessary, involving the integration of modern technologies, supportive policies, and collaborative efforts. Key strategies include:

1. Strengthening Market Infrastructure

Investment in Storage Facilities: Building modern storage facilities can help reduce post-harvest losses and ensure that produce remains fresh until it reaches the market.

Improving Transportation Networks: Enhancing transportation infrastructure can facilitate faster and more efficient movement of produce from farms to markets.

2. Leveraging Technology

Digital Platforms: Developing and promoting digital platforms can connect farmers directly with buyers, reducing dependency on middlemen and ensuring fair pricing.

Market Information Systems: Implementing market information systems can provide farmers with real-time data on prices, demand, and supply trends, enabling better decision-making.

3. Enhancing Farmer Training and Capacity Building

Extension Services: Strengthening agricultural extension services can provide farmers with the knowledge and skills needed to adopt advanced marketing practices and improve productivity.

Financial Literacy: Training programs focused on financial literacy can help farmers manage their finances better, access credit, and invest in improved marketing strategies.

4. Promoting Value Addition and Branding

Processing Facilities: Establishing local processing facilities can enable farmers to add value to their produce, increasing their income potential.

Brand Development: Supporting farmers in developing strong brands for their produce can enhance market differentiation and consumer loyalty.

5. Facilitating Access to Credit

Microfinance and Cooperative Models: Promoting microfinance and cooperative models can provide farmers with easier access to credit, enabling them to invest in better agricultural practices and marketing strategies.

Government Subsidies: Implementing targeted subsidies and financial support programs can help reduce the cost burden on farmers and encourage the adoption of advanced marketing techniques.

6. Strengthening Policy Frameworks

Consistent Policies: Developing consistent and supportive policies for agricultural marketing can create a stable environment for farmers to plan and operate effectively.

Public-Private Partnerships: Encouraging public-private partnerships can leverage resources and expertise to improve market infrastructure and services.

The Promise of Advanced Marketing

Advanced marketing offers several potential benefits to address these challenges:

1. **Increased Efficiency**: Modern marketing techniques and technologies can streamline the supply chain, reduce waste, and improve overall efficiency.
2. **Higher Profits**: By reducing dependency on middlemen and accessing wider markets, farmers can achieve better prices for their produce.
3. **Reduced Losses**: Improved logistics, better storage solutions, and precision agriculture can significantly reduce post-harvest losses.
4. **Market Intelligence**: Advanced marketing tools provide real-time market data, helping farmers make informed decisions regarding crop selection, pricing, and selling strategies.
5. **Enhanced Market Access**: Digital platforms and e-commerce solutions can connect farmers directly with consumers and larger markets, expanding their reach and opportunities.

Components of Advanced Marketing in Agriculture

Digital Platforms and E-Commerce

Online Marketplaces: Digital platforms like e-commerce websites and mobile applications provide a direct connection between farmers and consumers. Examples include platforms like Alibaba's 1688.com for bulk sales and Fresh Direct for direct-to-consumer sales.

Benefits

- **Wider Reach**: Farmers can access national and international markets.
- **Fair Pricing**: Direct sales reduce intermediary costs and increase farmer profits.
- **Convenience**: Consumers benefit from the convenience of online shopping.

Challenges

- **Digital Literacy**: Farmers need training to effectively use digital platforms.
- **Infrastructure**: Reliable internet and delivery logistics are essential for the success of digital platforms.

Precision Agriculture and Data Analytics

Precision Agriculture: This involves using technology to monitor and manage fields at a micro level, optimizing inputs like water, fertilizer, and pesticides to maximize yields and quality.

Data Analytics: Advanced data analytics can provide insights into market trends, consumer preferences, and optimal pricing strategies.

Benefits

- **Efficiency**: Better resource management leads to higher yields and reduced costs.
- **Informed Decisions**: Data-driven insights help in making strategic decisions about planting, harvesting, and marketing.

Challenges

- **Initial Investment**: High initial costs for technology and equipment.
- **Technical Skills**: Farmers need to acquire new skills to interpret data and use precision agriculture tools effectively.

Branding and Value Addition

Branding: Developing a strong brand identity helps differentiate produce in a crowded market. This can be particularly effective for organic or specialty products.

Value Addition: Processing raw agricultural produce into higher-value products, such as turning tomatoes into sauces or milk into cheese, can significantly increase profitability.

Benefits

- **Higher Prices**: Branded and value-added products can command premium prices.
- **Market Differentiation**: A strong brand can create a loyal customer base and reduce price competition.

Challenges

- **Marketing Costs**: Building a brand requires investment in marketing and packaging.
- **Quality Control**: Maintaining consistent quality is crucial for brand reputation.

Sustainable Practices and Certification

Sustainability: Adopting sustainable farming practices not only helps the environment but can also be a significant marketing point.

Certification: Certifications such as organic, fair trade, and non-GMO can add value and appeal to certain consumer segments.

Benefits

- **Market Appeal**: Growing consumer preference for sustainably produced goods can open new market opportunities.
- **Price Premiums**: Certified products often sell at higher prices.

Challenges

- **Certification Costs**: The process of obtaining and maintaining certifications can be expensive and time-consuming.
- **Compliance**: Farmers need to adhere to strict guidelines to maintain certification.

Successful Case Studies

1. India: Digital Green

Background: Digital Green is an initiative in India that uses video technology to disseminate agricultural information to farmers. It provides a platform where farmers can share their knowledge and best practices.

Implementation: Farmers are trained to create and share short videos on various agricultural practices, which are then shown in local community screenings.

Outcomes

- **Increased Adoption**: The participatory approach has led to high levels of adoption of new practices.
- **Cost-Effective**: The use of video technology is more cost-effective than traditional extension services.

- **Scalability**: The model is scalable and has been expanded to other countries.

2. Kenya: M-Farm

Background: M-Farm is a mobile platform in Kenya that provides market price information, group selling opportunities, and input supply to farmers.

Implementation: Farmers can access real-time price information for various markets through SMS. The platform also facilitates collective bargaining and bulk purchasing of inputs.

Outcomes

- **Increased Incomes**: Farmers receive better prices for their produce by avoiding middlemen.
- **Empowerment**: Access to information empowers farmers to make informed decisions.
- **Expanded Reach**: The mobile platform reaches farmers in remote areas.

3. United States: Local Roots

Background: Local Roots is an urban farming initiative in New York City that uses hydroponic technology to grow fresh produce in shipping containers.

Implementation: The produce is sold directly to consumers through subscription models and local markets.

Outcomes

- **Sustainability**: The hydroponic system uses less water and land compared to traditional farming.
- **Freshness**: Produce is delivered within hours of harvesting, ensuring maximum freshness.
- **Community Engagement**: The initiative promotes local food systems and engages urban communities.

4. Brazil: Coffee Cooperatives

Background: Coffee cooperatives in Brazil have successfully used branding and value addition to improve market access and farmer incomes.

Implementation: Cooperatives provide training on sustainable farming practices, help with certification processes, and market the coffee under a collective brand.

Outcomes

- **Market Access**: Cooperatives have accessed international markets and premium price segments.
- **Improved Practices**: Training and support have improved farming practices and yields.
- **Economic Benefits**: Farmers receive higher prices and more stable incomes through cooperative marketing.

Challenges in Implementing Advanced Marketing Strategies

1. Infrastructure and Technological Barriers

Digital Divide: Access to technology and internet connectivity is still limited in many rural areas, impeding the implementation of digital marketing solutions.

Logistics and Transportation: Efficient logistics and transportation networks are crucial for timely delivery of produce. Poor infrastructure can lead to delays and increased post-harvest losses.

2. Financial Constraints

Investment Needs: Advanced marketing strategies often require significant initial investments in technology, training, and infrastructure. Small-scale farmers may lack access to the necessary capital.

Risk Management: Transitioning to advanced marketing methods can involve risks, such as fluctuating market prices and changing consumer preferences. Farmers need support in managing these risks.

3. Training and Capacity Building

Skill Development: Farmers and extension workers need training to effectively use new technologies and marketing strategies. This requires continuous capacity-building efforts.

Knowledge Transfer: Effective knowledge transfer mechanisms are needed to ensure that farmers can adopt and benefit from advanced marketing techniques.

4. Market Dynamics

Consumer Awareness: Educating consumers about the benefits of advanced and sustainable farming practices is essential to create demand for these products.

Market Access: Ensuring that small-scale farmers can access larger and more lucrative markets requires coordinated efforts and supportive policies.

Strategies to Overcome Challenges

1. Government and Policy Support

Incentives and Subsidies: Governments can provide financial incentives and subsidies to support the adoption of advanced marketing strategies and technologies.

Infrastructure Development: Investing in rural infrastructure, including roads, internet connectivity, and logistics, can facilitate the implementation of advanced marketing solutions.

Regulatory Frameworks: Creating favourable regulatory frameworks for certification, e-commerce, and digital platforms can encourage innovation and growth in agricultural marketing.

2. Partnerships and Collaborations

Public-Private Partnerships: Collaborations between governments, private sector companies, and non-profit organizations can leverage resources and expertise to support advanced marketing initiatives.

Farmer Cooperatives: Encouraging the formation of cooperatives can help small-scale farmers pool resources, access markets, and achieve economies of scale.

Research and Development: Investing in R&D to develop new technologies and practices tailored to local conditions can enhance the effectiveness of advanced marketing strategies.

3. Capacity Building and Education

Training Programs: Establishing training programs for farmers and extension workers on advanced marketing techniques and technologies is crucial for successful implementation.

Demonstration Projects: Implementing demonstration projects can showcase the benefits of advanced marketing strategies and encourage wider adoption.

Knowledge Sharing: Creating platforms for knowledge sharing and collaboration among farmers, researchers, and market actors can foster innovation and best practices.

4. Leveraging Technology

Mobile Solutions: Developing mobile-based solutions for market information, transactions, and communication can bridge the digital divide and reach remote farmers.

Blockchain and Traceability: Using blockchain technology for traceability can enhance transparency and build consumer trust in the supply chain.

Data Analytics: Utilizing data analytics to provide real-time market insights and predictive analytics can help farmers make informed decisions and optimize their marketing strategies.

Conclusion

Advanced marketing of agricultural produce is essential for modernizing the agricultural sector, improving market efficiency, and enhancing farmer livelihoods. By integrating digital platforms, precision agriculture, branding, value addition, and sustainable practices, farmers can better connect with markets, reduce post-harvest losses, and achieve higher economic returns. While challenges such as infrastructure, financial constraints, and training needs remain, coordinated efforts from governments, private sector actors, and non-profit organizations can overcome these obstacles. By investing in advanced marketing strategies and technologies, the agricultural sector can become more resilient, sustainable, and profitable, benefiting farmers and consumers alike.

References

Hart, P. (2014). Marketing agricultural produce. In The Geography of Agriculture in Developed Market Economies (pp. 162-206). Routledge.

Hingley, M., & Lindgreen, A. (2002). Marketing of agricultural products: case findings. British food journal, 104(10), 806-827.

Kimavath, N. Suggestions to Improve Agricultural Marketing.

Kohls, R. L., & Uhl, J. N. (2002). Marketing of agricultural products (No. Ed. 9). Prentice-Hall Inc..

Kulikov, I. M., & Minakov, I. A. (2020). Commercial activity of agricultural producers. Journal of Advanced Research in Law and Economics, 11(3 (49)), 913-920.

Sharma, M., & Patil, C. (2018). Recent trends and advancements in agricultural research: An overview. Journal of Pharmacognosy and Phytochemistry, 7(2), 1906-1910.

19

Advanced Storage Methods of Agricultural Crops

***Anushka Singh*[1], *Chetan Lal*[2], *Shradha Parmar*[3] *and Durgesh Kumar Maurya*[4]**

[1]*Department of Horticulture, K.P.H.E.I. Affiliated by Prof. Rajendra Singh (Rajju Bhaiya) University, Prayagraj, Uttar Pradesh*
[2]*Processing and Food Engineering, SKUAST, Kashmir, Jammu and Kashmir*
[3]*Rajmata Vijayaraje Scindia Krishi Vishwa Vidyalaya, Gwalior, Madhya Pradesh*
[4]*Department of Agronomy, CSAUAT, Kanpur, Uttar Pradesh*

Abstract

Advanced storage methods of agricultural crops play a pivotal role in mitigating post-harvest losses, ensuring food security, and enhancing the economic viability of farming. These methods encompass a range of innovative techniques and technologies designed to prolong the shelf life of crops, maintain quality, and reduce spoilage. From controlled atmosphere storage and modified atmosphere packaging to cold storage facilities and hermetic storage bags, advanced storage methods offer diverse solutions tailored to different crops and environments. This chapter explores the key principles and benefits of advanced storage methods, highlights notable technologies and approaches, and underscores their significance in addressing the challenges of post-harvest management in agriculture.

Keywords: *atmosphere, hermetic, technologies, storage, principles*

Introduction

Post-harvest losses of agricultural crops have long been a significant challenge facing farmers worldwide. According to the Food and Agriculture Organization (FAO) of the United Nations, around one-third of all food produced for human consumption is lost or wasted each year. Advanced storage methods offer innovative solutions to mitigate these losses, ensuring food security, enhancing economic viability, and promoting sustainability in agriculture. This comprehensive article delves into the principles, benefits, technologies, and challenges of advanced storage methods for agricultural crops, providing insights into their role in modern agriculture and their potential to revolutionize post-harvest management.

The Importance of Advanced Storage Methods

Advanced storage methods play a pivotal role in modern agriculture, offering innovative solutions to address the challenges of post-harvest losses, ensure food security, and promote sustainability. The importance of advanced storage methods can be understood through various key aspects:

1. **Minimizing Post-Harvest Losses:** Post-harvest losses are a significant challenge in agriculture, resulting from factors such as inadequate storage facilities, pests, diseases, and environmental conditions. Advanced storage methods provide optimal conditions for crop storage, including temperature, humidity, and gas composition, thereby minimizing spoilage and preserving the quality of agricultural produce.
2. **Ensuring Food Security:** Efficient storage methods contribute to food security by preserving the quality and nutritional value of agricultural crops. By extending the shelf life of perishable crops, advanced storage methods ensure a stable food supply throughout the year, especially in regions prone to seasonal shortages or disruptions. This contributes to food security at both the local and global levels.
3. **Enhancing Economic Viability:** Post-harvest losses represent a significant economic burden for farmers, leading to reduced incomes and wasted resources. Advanced storage methods enable farmers to store their crops for longer periods, allowing them to sell at higher prices during peak demand seasons and avoid selling at distress prices immediately after harvest. This enhances the economic viability of farming operations and improves the livelihoods of farmers.
4. **Promoting Sustainability:** Reducing post-harvest losses through advanced storage methods has environmental benefits, such as reducing greenhouse gas emissions associated with food waste and conserving natural resources used in crop production. By promoting sustainable practices, advanced storage methods contribute to the overall sustainability of agriculture, aligning with global efforts to achieve food security while minimizing environmental impact.

5. **Facilitating Market Access:** Advanced storage methods enable farmers to store their produce for longer periods without compromising quality, allowing them to access distant and lucrative markets. By extending the market reach of agricultural produce, advanced storage methods create new opportunities for farmers to increase their income and expand their businesses. This contributes to the overall economic development of rural communities.

In conclusion, advanced storage methods are indispensable tools for modern agriculture, offering a range of benefits that contribute to food security, economic viability, and sustainability. By minimizing post-harvest losses, ensuring food quality, and facilitating market access, advanced storage methods play a crucial role in improving the efficiency and resilience of agricultural systems, ultimately benefiting farmers, consumers, and the environment.

Principles of Advanced Storage Methods

The principles of advanced storage methods in agriculture revolve around creating optimal conditions to prolong the shelf life of agricultural crops, minimize post-harvest losses, and maintain product quality. These principles encompass various factors, including temperature control, humidity management, gas composition regulation, and pest and disease management. Below are the key principles of advanced storage methods:

1. Temperature Control

Temperature regulation is fundamental to preserving the quality and freshness of agricultural crops during storage. Maintaining the appropriate temperature slows down biochemical processes, such as respiration and microbial growth, which lead to spoilage. The principle of temperature control involves

- **Cooling Systems**: Utilizing refrigeration units, cold storage facilities, or cooling technologies to maintain low temperatures.
- **Temperature Monitoring**: Regularly monitoring and adjusting storage temperatures to ensure they remain within optimal ranges for specific crops.

2. Humidity Management

Humidity levels play a crucial role in preventing moisture-related deterioration and preserving the texture and appearance of agricultural crops. The principle of humidity management involves

- **Dehumidification**: Using ventilation systems, desiccants, or humidity controllers to remove excess moisture from the storage environment.
- **Humidity Monitoring**: Monitoring humidity levels and preventing conditions that promote mold growth, caking, or dehydration of crops.

3. Gas Composition Control

Controlling the composition of gases surrounding agricultural crops can significantly influence their shelf life and quality. The principle of gas composition control involves:

- **Modified Atmosphere Packaging (MAP)**: Packaging crops in controlled atmospheres with modified levels of oxygen, carbon dioxide, and ethylene to slow down ripening and inhibit microbial growth.
- **Controlled Atmosphere Storage (CAS)**: Creating storage environments with precise gas compositions to maintain product quality and extend shelf life.

4. Pest and Disease Management

Preventing infestations and diseases is essential for preserving the quality and marketability of agricultural crops during storage. The principle of pest and disease management involves

- **Hermetic Storage**: Using airtight containers or bags to suffocate pests and inhibit their reproduction without the need for chemical pesticides.
- **Integrated Pest Management (IPM)**: Implementing holistic approaches that combine biological, cultural, and chemical control methods to manage pests and diseases effectively.

5. Quality Monitoring and Control

Regular monitoring of storage conditions and product quality is essential to identify and address any issues that may arise during storage. The principle of quality monitoring and control involves

- **Regular Inspections**: Conducting visual inspections and quality assessments to identify signs of spoilage, damage, or deterioration.
- **Quality Assurance Practices**: Implementing hygiene protocols, sanitation measures, and quality control procedures to maintain product integrity and safety.

6. Energy Efficiency and Sustainability

Promoting energy-efficient and sustainable practices is crucial for reducing environmental impact and operational costs associated with advanced storage methods. The principle of energy efficiency and sustainability involves:

- **Optimized Operations**: Implementing energy-efficient technologies, such as insulation, LED lighting, and renewable energy sources, to minimize energy consumption.
- **Resource Conservation**: Adopting sustainable practices, such as recycling, waste reduction, and water conservation, to minimize environmental footprint and promote long-term sustainability.

By adhering to these principles, farmers and storage facility operators can maximize the effectiveness of advanced storage methods, minimize post-harvest losses, and ensure the delivery of high-quality agricultural crops to markets and consumers.

Advanced Storage Technologies and Innovations

Advanced storage technologies and innovations have revolutionized the way agricultural crops are preserved, extending their shelf life, maintaining quality, and reducing post-harvest losses. These technologies leverage cutting-edge science and engineering to create optimal storage environments tailored to the specific needs of different crops. Below are some of the most impactful advanced storage technologies and innovations in agriculture:

1. Controlled Atmosphere Storage (CAS)

Controlled atmosphere storage involves modifying the composition of gases surrounding stored crops to create optimal conditions for preservation. Key innovations in CAS include

- **Gas Monitoring and Control Systems**: Automated systems that continuously monitor and adjust oxygen, carbon dioxide, and ethylene levels in storage chambers to maintain desired conditions.
- **Dynamic Control Algorithms**: Advanced algorithms that optimize gas compositions based on crop type, maturity stage, and environmental factors to maximize shelf life and quality.

2. Modified Atmosphere Packaging (MAP)

Modified atmosphere packaging involves packaging agricultural crops in specialized materials with modified gas compositions to extend shelf life. Innovations in MAP include:

- **Active Packaging Systems**: Packaging materials embedded with oxygen scavengers, carbon dioxide emitters, and ethylene absorbers to actively regulate gas levels and preserve product freshness.
- **Smart Packaging Technologies**: Intelligent packaging solutions equipped with sensors and indicators that monitor product quality and provide real-time feedback on storage conditions.

3. Cold Storage Facilities

Cold storage facilities utilize refrigeration systems to maintain low temperatures and preserve the quality of perishable crops. Innovations in cold storage include

- **Energy-Efficient Refrigeration Systems**: High-efficiency refrigeration units, cold room insulation, and temperature management systems that minimize energy consumption and reduce operational costs.
- **Cold Chain Management Systems**: Integrated cold chain solutions that track and monitor product temperatures throughout the supply chain, ensuring consistent cold storage conditions from farm to market.

4. Hermetic Storage Bags

Hermetic storage bags create an airtight environment that suffocates pests and inhibits microbial growth, preserving the quality of stored crops without the need for chemical pesticides. Innovations in hermetic storage bags include

- **Multi-Layered Barrier Films**: Advanced polymer films with superior barrier properties that prevent gas exchange and moisture ingress, ensuring long-term storage stability.
- **Biodegradable Materials**: Environmentally friendly hermetic bags made from biodegradable materials that reduce plastic waste and promote sustainability.

5. Vacuum Cooling

Vacuum cooling rapidly reduces the temperature of freshly harvested crops by removing air and moisture from the produce. Innovations in vacuum cooling include

- **High-Speed Vacuum Chambers**: Advanced vacuum chambers equipped with rapid cooling cycles that minimize processing time and preserve product quality.

- **Adaptive Cooling Algorithms**: Intelligent cooling systems that adjust vacuum pressure and cooling rates based on crop characteristics and ambient conditions to optimize cooling efficiency.

6. Smart Storage Solutions

Smart storage solutions leverage sensors, data analytics, and automation to monitor and control storage conditions in real-time. Innovations in smart storage include

- **Wireless Sensor Networks**: Wireless sensor networks deployed throughout storage facilities to monitor temperature, humidity, gas levels, and other critical parameters.
- **Cloud-Based Analytics Platforms**: Cloud-based analytics platforms that collect and analyse sensor data to identify trends, predict storage failures, and optimize storage settings for maximum efficiency.

7. Nanotechnology Applications

Nanotechnology applications in storage involve the use of nanostructured materials and coatings to enhance the properties of packaging materials and storage containers. Innovations in nanotechnology include

- **Nano-Enhanced Packaging Films**: Nanocomposite films with antimicrobial properties, gas barrier capabilities, and mechanical strength that improve the shelf life and safety of packaged crops.
- **Nano-Enabled Smart Sensors**: Miniaturized smart sensors embedded in packaging materials that provide real-time information on product freshness, contamination, and storage conditions.

These advanced storage technologies and innovations represent the forefront of post-harvest management in agriculture, offering unprecedented opportunities to reduce waste, improve efficiency, and ensure the availability of high-quality food products for consumers around the world.

Challenges of Implementing Advanced Storage Methods

Implementing advanced storage methods in agriculture is not without its challenges. Despite the significant benefits they offer, various factors can impede their adoption and effectiveness. Here are some of the key challenges:

1. **Initial Investment Costs:** One of the primary challenges is the high upfront costs associated with implementing advanced storage methods. Purchasing and installing equipment such as controlled atmosphere storage chambers, refrigeration units, and specialized packaging

materials require substantial investment, which may be prohibitive for small-scale farmers or resource-constrained regions.

2. **Technical Expertise:** Operating and maintaining advanced storage systems often requires specialized knowledge and skills. Many farmers and storage facility operators may lack the technical expertise to use complex equipment effectively, leading to suboptimal performance and potential failures. Training programs and technical support are essential to address this challenge.

3. **Energy Consumption:** Advanced storage methods, such as cold storage facilities and controlled atmosphere systems, consume significant amounts of energy to maintain optimal storage conditions. High energy costs and reliance on fossil fuels contribute to environmental degradation and increase operational expenses, posing challenges for sustainable implementation.

4. **Infrastructure Limitations:** Inadequate infrastructure, including unreliable electricity supply, poor transportation networks, and lack of storage facilities, hinders the implementation of advanced storage methods in rural and remote areas. Improving infrastructure, such as access to electricity and roads, is essential to expand the reach of advanced storage technologies.

5. **Market Access:** Limited market access and distribution networks constrain the adoption of advanced storage methods, as farmers may struggle to sell their preserved crops in local and regional markets. Strengthening market linkages, developing value chains, and improving transportation logistics are critical to realizing the benefits of advanced storage.

6. **Knowledge and Awareness:** Many farmers may lack awareness of the benefits and proper techniques for implementing advanced storage methods. Knowledge dissemination, training programs, and extension services are essential to educate farmers about the importance of post-harvest management and provide guidance on adopting advanced storage technologies.

7. **Maintenance and Support:** Ensuring the long-term viability of advanced storage systems requires regular maintenance and technical support. However, access to maintenance services and spare parts may be limited in rural areas, leading to operational challenges and downtime. Establishing support networks and service providers is crucial to address maintenance needs.

8. **Climate and Environmental Factors:** Extreme weather events, such as storms, floods, and heatwaves, can pose risks to advanced storage facilities and disrupt supply chains. Climate variability and environmental factors may also affect the performance of storage systems, requiring adaptation strategies and contingency plans to mitigate risks.

9. **Regulatory and Policy Constraints:** Inconsistent regulations, import/export policies, and market standards may create barriers to the adoption and trade of advanced storage technologies. Harmonizing regulations, providing incentives for investment, and fostering public-private partnerships can help overcome regulatory challenges and promote innovation in post-harvest management.

Addressing these challenges requires a coordinated effort involving policymakers, industry stakeholders, research institutions, and development agencies. By investing in infrastructure, providing technical support, raising awareness, and fostering enabling policies, the adoption of advanced storage methods can be facilitated, leading to improved food security, reduced post-harvest losses, and enhanced livelihoods for farmers.

Case Studies and Success Stories

Case studies and success stories provide valuable insights into the real-world application and impact of advanced storage methods in agriculture. Here are some examples

1. **Zero Energy Cool Chamber (ZECC) in India:** The Zero Energy Cool Chamber (ZECC) is a low-cost, energy-efficient storage solution developed by researchers in India to preserve fruits and vegetables. The ZECC utilizes passive cooling techniques, such as evaporative cooling and thermal insulation, to maintain low temperatures without electricity. By storing produce in the ZECC, farmers can extend the shelf life of their crops and reduce post-harvest losses, especially during hot and dry seasons. The ZECC has been successfully adopted by small-scale farmers in rural areas, enabling them to access distant markets and obtain better prices for their produce.

2. **Purdue Improved Crop Storage (PICS) Bags in Africa:** Purdue Improved Crop Storage (PICS) bags are hermetic storage bags developed by Purdue University to protect grains from insect pests and mold growth. PICS bags create an airtight environment that suffocates pests and inhibits their reproduction, preventing losses due to infestations. These bags have been widely adopted by smallholder farmers in Africa,

where post-harvest losses are a significant challenge. By using PICS bags, farmers can store their grains for longer periods without the need for chemical pesticides, resulting in higher quality and better market prices.

3. **Controlled Atmosphere Storage (CAS) in the United States:** Large-scale fruit producers in the United States utilize controlled atmosphere storage systems to extend the shelf life of apples, pears, and other fruits. CAS systems regulate oxygen, carbon dioxide, and ethylene levels within storage chambers to create optimal conditions for preservation. By reducing respiration rates and inhibiting microbial growth, CAS systems preserve fruit quality during long-term storage and shipping. This allows producers to supply markets year-round and meet consumer demand for fresh fruit even during off-seasons. CAS has become a standard practice in the fruit industry, contributing to increased profitability and market competitiveness.
4. **Solar-Powered Cold Storage in Bangladesh:** In Bangladesh, solar-powered cold storage facilities have been implemented to address the challenges of post-harvest losses and lack of access to electricity in rural areas. These facilities use solar panels to generate electricity for refrigeration units, allowing farmers to store perishable crops such as vegetables, fish, and dairy products. By utilizing renewable energy sources, solar-powered cold storage facilities reduce operating costs and environmental impact while improving food security and income opportunities for farmers. This innovative approach has been replicated in other countries facing similar challenges.
5. **Vacuum Cooling in Europe:** In Europe, vacuum cooling technology is widely used to rapidly reduce the temperature of freshly harvested produce, such as leafy greens and berries. Vacuum cooling removes air and moisture from the produce, causing evaporative cooling and lowering the temperature within minutes. This process preserves the freshness and quality of the produce, extending its shelf life and reducing post-harvest losses. Vacuum cooling is commonly employed in the fresh produce industry, allowing farmers to meet strict quality standards and supply perishable products to distant markets with minimal spoilage.

These case studies and success stories demonstrate the effectiveness of advanced storage methods in improving post-harvest management, enhancing food security, and increasing income opportunities for farmers. By leveraging innovative technologies and sustainable practices, advanced storage methods contribute to a more efficient and resilient agricultural system, benefiting both producers and consumers.

Conclusion

Advanced storage methods of agricultural crops are indispensable tools for modernizing post-harvest management, ensuring food security, and promoting sustainability in agriculture. By leveraging innovative technologies and practices, farmers can minimize losses, preserve quality, and enhance market access, ultimately improving livelihoods and reducing environmental impact. While challenges remain, continued investment in research, infrastructure, and capacity-building initiatives is essential for realizing the full potential of advanced storage methods and building

References

Amit, S.K., Uddin, M.M., Rahman, R., Islam, S.R. and Khan, M.S. (2017). A review on mechanisms and commercial aspects of food preservation and processing. Agriculture & Food Security, 6, 1-22.

Kumar, D., & Kalita, P. (2017). Reducing postharvest losses during storage of grain crops to strengthen food security in developing countries. Foods, 6(1), 8.

Mahmood, M. H., Sultan, M., & Miyazaki, T. (2019). Significance of temperature and humidity control for agricultural products storage: overview of conventional and advanced options. International Journal of Food Engineering, 15(10), 20190063.

Navarro, S. (2012). Advanced grain storage methods for quality preservation and insect control based on aerated or hermetic storage and IPM. Journal of Agricultural Engineering, 49(1), 13-20.

Pekmez, H. (2016). Cereal storage techniques: A review. J. Agric. Sci. Technol. B, 6(2), 67-71.

Valdramidis, V. P., & Koutsoumanis, K. P. (2016). Challenges and perspectives of advanced technologies in processing, distribution and storage for improving food safety. Current Opinion in Food Science, 12, 63-69.

20

Post-Harvest Management of Agricultural Crops

Ashish Ashok Patil and Ajaykumar Singh

Department of Processing and Food Engineering, Sam Higginbottom University of Agriculture Technology and Sciences, Prayagraj Uttar Pradesh

Abstract

Post-harvest management of agricultural crops is a critical aspect of the food supply chain, encompassing a range of activities aimed at preserving the quality and extending the shelf life of harvested produce. The abstract on post-harvest management highlights its significance in ensuring food security, reducing post-harvest losses, and enhancing economic viability for farmers. It emphasizes the multifaceted nature of post-harvest management, which involves handling, storage, processing, and distribution of agricultural crops. The abstract may also touch upon key challenges faced in post-harvest management, such as inadequate storage facilities, pest infestations, and market access constraints. Overall, the abstract provides a concise overview of the importance and complexity of post-harvest management in sustaining agriculture and meeting the nutritional needs of populations globally.

Keywords: *management, critical, viability, pest, concise*

Introduction

Post-harvest management of agricultural crops is a crucial component of the food supply chain, encompassing activities from harvesting to consumption aimed at preserving the quality, safety, and nutritional value of harvested produce. Effective post-harvest management not only reduces losses but also contributes to food security, economic sustainability, and environmental conservation. This comprehensive article delves into the various aspects of post-harvest management, including harvesting techniques, storage methods, processing technologies, and distribution strategies, highlighting their importance in ensuring food availability, improving livelihoods, and mitigating environmental impact.

Importance of Post-Harvest Management

Post-harvest management is a critical aspect of agricultural production with far-reaching implications for food security, economic sustainability, and environmental conservation. The importance of post-harvest management can be understood through several key aspects:

1. **Minimizing Food Losses and Waste:** Post-harvest losses, including spoilage, damage, and decay, result in significant quantities of food being wasted globally. Effective post-harvest management practices help minimize these losses by preserving the quality and extending the shelf life of harvested crops. By reducing food waste, post-harvest management contributes to the efficient utilization of resources and supports efforts to address hunger and malnutrition.
2. **Ensuring Food Security:** Agricultural productivity alone is not sufficient to guarantee food security; efficient post-harvest management is equally crucial. By maintaining the quality and availability of harvested crops, post-harvest management helps ensure a stable and reliable food supply chain. This is particularly important in regions prone to seasonal fluctuations, environmental hazards, and market disruptions, where effective post-harvest management can help buffer against food shortages and crises.
3. **Enhancing Economic Viability:** Post-harvest losses represent a direct economic loss to farmers, producers, and other stakeholders along the food supply chain. Effective post-harvest management practices, such as proper handling, storage, and processing techniques, help preserve the value of agricultural produce and maximize returns on investment. By minimizing losses and optimizing marketable yields, post-harvest management enhances the economic viability of agricultural enterprises and contributes to rural livelihoods and incomes.
4. **Promoting Sustainable Agriculture:** Efficient post-harvest management practices contribute to the sustainability of agriculture by reducing resource wastage, environmental impact, and greenhouse gas emissions. By optimizing storage conditions, minimizing energy consumption, and adopting eco-friendly packaging materials, post-harvest management supports environmentally sustainable food production systems. Additionally, by reducing the need for additional

land, water, and inputs to compensate for losses, effective post-harvest management helps conserve natural resources and mitigate the environmental footprint of agriculture.

5. **Supporting Market Access and Trade:** High-quality post-harvest management practices are essential for accessing domestic and international markets and meeting stringent quality and safety standards. By ensuring that crops reach consumers in optimal condition, post-harvest management facilitates market access, enhances competitiveness, and expands trade opportunities for agricultural producers. This, in turn, stimulates economic growth, fosters rural development, and strengthens food systems resilience to external shocks and disruptions.

In summary, post-harvest management plays a pivotal role in ensuring food security, enhancing economic viability, promoting sustainability, and facilitating market access in agriculture. By adopting efficient post-harvest management practices and investing in infrastructure, technology, and capacity building, stakeholders can maximize the value of agricultural produce, minimize losses, and contribute to a more resilient and sustainable food system.

Harvesting Techniques

Harvesting techniques are essential practices employed in agriculture to ensure the timely and efficient collection of crops at the peak of their maturity. These techniques vary depending on the type of crop, its intended use, and environmental conditions. Here are some common harvesting techniques

1. **Timing and Maturity:** Harvesting at the right time and maturity stage is crucial to maximize yield and quality. Different crops have specific indicators of maturity, such as colour, size, texture, and flavour, which farmers must carefully monitor. Harvesting too early can result in underdeveloped crops with poor flavour and nutritional content, while harvesting too late may lead to overripe or damaged produce.

2. **Manual Harvesting:** Manual harvesting involves handpicking or cutting crops using handheld tools such as knives, sickles, or shears. This technique is commonly used for fruits, vegetables, and flowers that require careful handling to avoid bruising or damage. Manual harvesting allows farmers to selectively harvest ripe crops while leaving unripe ones to continue maturing.

3. **Mechanical Harvesting:** Mechanical harvesting utilizes specialized equipment such as combine harvesters, reapers, and threshers to automate the harvesting process. This technique is suitable for large-

scale cultivation of grains, cereals, and oilseeds, where efficiency and speed are essential. Mechanical harvesters can significantly reduce labour requirements and harvest time, increasing productivity and minimizing post-harvest losses.

4. **Selective Harvesting:** Selective harvesting involves harvesting only mature or marketable portions of a crop while leaving the rest to continue growing. This technique is common in perennial crops such as fruits, nuts, and berries, where not all fruits reach maturity simultaneously. Selective harvesting allows farmers to optimize yield and quality by harvesting ripe fruits while allowing others to ripen further.

5. **Continuous Harvesting:** Continuous harvesting involves harvesting crops over an extended period rather than all at once. This technique is common in crops with extended harvesting seasons, such as leafy greens, herbs, and certain vegetables. Continuous harvesting allows farmers to stagger production, maintain a steady supply to markets, and extend the income-generating period.

6. **Post-Harvest Handling:** Post-harvest handling practices, such as sorting, grading, and packaging, are essential components of harvesting techniques. Sorting involves separating crops based on size, colour, and quality, while grading assigns them into different categories based on predetermined standards. Packaging protects harvested crops from physical damage, contamination, and spoilage during storage and transportation.

7. **Consideration of Environmental Factors:** Environmental factors such as weather conditions, temperature, and humidity can influence harvesting decisions. Harvesting during favourable weather conditions, such as cool mornings or dry periods, can help preserve the quality and shelf life of harvested crops. Additionally, avoiding harvesting during periods of high humidity or rainfall can minimize the risk of mold growth and spoilage.

Storage Methods

Storage methods are crucial in agriculture for preserving the quality, freshness, and nutritional value of harvested crops over extended periods. Effective storage methods help minimize post-harvest losses, maintain marketability, and ensure a steady food supply. Here are some common storage methods used in agriculture:

1. **Cold Storage:** Cold storage involves storing crops at low temperatures to slow down metabolic processes and inhibit microbial growth, thereby extending shelf life. Cold storage facilities, such as refrigerated warehouses, walk-in coolers, and refrigerated trucks, maintain temperatures typically between 0°C to 10°C (32°F to 50°F), depending on the specific requirements of the crop. Cold storage is particularly suitable for perishable crops like fruits, vegetables, dairy products, and meat.
2. **Controlled Atmosphere Storage (CAS):** Controlled atmosphere storage involves modifying the composition of gases, including oxygen, carbon dioxide, and ethylene, within storage chambers to create optimal conditions for crop preservation. By controlling gas levels, CAS inhibits ripening, slows down decay, and extends the storage life of fruits, vegetables, and grains. CAS systems are commonly used for long-term storage of apples, pears, potatoes, and other produce.
3. **Modified Atmosphere Packaging (MAP):** Modified atmosphere packaging involves packaging agricultural crops in specialized materials with modified gas compositions to create an optimal atmosphere that slows down deterioration. MAP extends the shelf life of packaged foods, maintains freshness, and reduces the need for chemical preservatives. MAP is commonly used for fruits, vegetables, bakery products, and processed meats.
4. **Drying:** Drying is a traditional method of preserving crops by reducing their moisture content to inhibit microbial growth and prevent spoilage. Various drying techniques, including sun drying, mechanical drying, and freeze-drying, are employed depending on the crop type and intended use. Dried fruits, vegetables, herbs, and grains have an extended shelf life and can be stored for long periods without refrigeration.
5. **Canning and Preservation:** Canning and preservation involve heat processing and sealing of crops in airtight containers to prevent spoilage and contamination. Canned fruits, vegetables, jams, and pickles retain their flavour, texture, and nutritional value for extended periods and can be stored at room temperature. Preservation methods, such as pickling, fermenting, and smoking, are also used to extend the shelf life of crops and enhance their flavour.
6. **Root Cellars and Silos:** Root cellars and silos are traditional storage structures used for storing root vegetables, grains, and fodder crops. Root cellars provide cool, dark, and humid conditions that are ideal for

preserving root crops like potatoes, carrots, and onions. Silos are tall, cylindrical structures used for storing grains, seeds, and animal feed in bulk quantities. Both root cellars and silos protect crops from pests, rodents, and environmental fluctuations.

7. **Hermetic Storage:** Hermetic storage involves sealing crops in airtight containers or bags to create an oxygen-deprived environment that suffocates pests and inhibits microbial growth. Hermetic storage is particularly effective for grains, pulses, and seeds, preventing infestations and maintaining quality without the need for chemical pesticides. Hermetic bags are portable, cost-effective, and suitable for small-scale farmers in remote areas.

8. **Smart Storage Solutions:** Smart storage solutions leverage sensors, data analytics, and automation to monitor and control storage conditions in real-time. Wireless sensor networks, cloud-based platforms, and smart packaging materials provide insights into temperature, humidity, gas levels, and product quality. Smart storage solutions optimize storage parameters, reduce energy consumption, and minimize losses, ensuring the delivery of high-quality crops to markets and consumers.

In conclusion, employing appropriate storage methods is essential for maintaining the quality, safety, and marketability of agricultural crops throughout the post-harvest period. By selecting suitable storage techniques, implementing proper handling practices, and investing in infrastructure and technology, farmers can minimize losses, optimize resource utilization, and ensure a sustainable food supply chain.

Processing Technologies

Processing technologies play a crucial role in transforming raw agricultural commodities into value-added products with enhanced shelf life, convenience, and marketability. These technologies involve various methods and techniques aimed at preserving the quality, flavour, and nutritional value of agricultural crops while meeting consumer preferences and industry standards. Here are some common processing technologies used in agriculture

1. **Cleaning and Sorting:** Cleaning and sorting are the initial steps in processing agricultural crops, involving the removal of impurities, foreign materials, and defective pieces. Cleaning may include washing, scrubbing, or air blowing to eliminate dirt, debris, and contaminants. Sorting entails separating crops based on size, shape, colour, and quality using mechanical or optical sorting equipment.

2. **Grading and Sizing:** Grading and sizing are processes that categorize agricultural crops into different quality grades and size categories based on predetermined standards. Grading ensures consistency and uniformity in product quality, appearance, and packaging, facilitating market access and consumer acceptance. Sizing equipment, such as screens, sieves, and rollers, are used to separate crops into various size fractions.

3. **Cutting and Chopping:** Cutting and chopping involve reducing the size of agricultural crops into smaller pieces or slices to facilitate handling, cooking, and consumption. Various cutting equipment, including knives, slicers, and choppers, are used to achieve desired shapes and sizes. Cutting and chopping are common in processing fruits, vegetables, meat, and poultry for further processing or packaging.

4. **Drying and Dehydration:** Drying and dehydration are methods of removing moisture from agricultural crops to inhibit microbial growth, reduce spoilage, and extend shelf life. Traditional sun drying, mechanical drying, and freeze-drying techniques are employed to preserve fruits, vegetables, herbs, and grains. Dehydrated products retain their nutritional value, flavour, and colour while offering increased storage stability and convenience.

5. **Freezing and Refrigeration:** Freezing and refrigeration are cold storage methods that preserve the freshness and quality of agricultural crops by maintaining low temperatures. Quick freezing techniques, such as blast freezing and cryogenic freezing, are used to rapidly freeze fruits, vegetables, seafood, and meat, locking in nutrients and flavours. Refrigeration extends the shelf life of perishable products and minimizes microbial growth, ensuring product safety and integrity.

6. **Canning and Preserving:** Canning and preserving involve heat processing and sealing agricultural crops in airtight containers to prevent spoilage and contamination. Canned fruits, vegetables, jams, and pickles retain their flavour, texture, and nutritional value for extended periods and can be stored at room temperature. Preservation methods, such as pickling, fermenting, and smoking, are also used to enhance flavour and extend shelf life.

7. **Fermentation and Culturing:** Fermentation and culturing are microbial processes that convert carbohydrates and sugars in agricultural crops into organic acids, alcohols, and gases under controlled conditions. Fermented products, such as yogurt, cheese, bread, and sauerkraut, undergo biochemical changes that enhance flavor, texture, and nutritional

value. Fermentation also improves digestibility and shelf life while inhibiting spoilage and pathogen growth.

8. **Milling and Grinding:** Milling and grinding are mechanical processes that reduce agricultural crops into fine powders, flours, or meal for various applications. Grain mills, hammer mills, and pulverisers are used to grind grains, seeds, and spices into fine particles suitable for food preparation, baking, and cooking. Milling and grinding enhance the digestibility, texture, and culinary versatility of agricultural products.

9. **Extrusion and Texturization:** Extrusion and texturization are advanced processing techniques used to transform agricultural crops into textured proteins, meat analogy, and snack foods. Extrusion involves forcing a mixture of ingredients through a die under high temperature and pressure to create textured shapes and structures. Texturized products offer meat-like texture, mouthfeel, and nutritional properties, appealing to vegetarian and health-conscious consumers.

10. **Packaging and Labelling:** Packaging and labelling are essential steps in processing agricultural products, ensuring their protection, preservation, and presentation to consumers. Packaging materials, such as bags, boxes, cans, and pouches, provide physical protection, barrier properties, and tamper-evident features. Labels convey product information, nutritional facts, ingredients, and branding elements to consumers, facilitating informed purchasing decisions.

In conclusion, processing technologies play a vital role in adding value to agricultural crops, diversifying product offerings, and meeting consumer demand for safe, nutritious, and convenient food products. By employing appropriate processing methods, adhering to quality standards, and embracing innovation, stakeholders in the food industry can enhance the competitiveness, sustainability, and profitability of agricultural value chains.

Distribution Strategies

1. **Cold Chain Logistics:** Maintaining a continuous cold chain from the point of harvest to the point of consumption is essential for preserving the quality and safety of perishable crops. Cold chain logistics involve proper handling, storage, and transportation of chilled or frozen products to prevent temperature fluctuations and ensure product integrity.

2. **Market Access:** Efficient distribution networks and market linkages are critical for connecting farmers to consumers and facilitating the flow of agricultural products from surplus to deficit regions. Access to markets

enables farmers to sell their produce at fair prices, reducing post-harvest losses and improving income opportunities.

3. **Value Addition:** Value-added processing and packaging activities, such as juicing, canning, and packaging, enhance the marketability and value of agricultural crops. Value-added products command higher prices and create new revenue streams for farmers, processors, and retailers, thereby increasing the economic returns from post-harvest activities.

Challenges in Post-Harvest Management

Post-harvest management faces several challenges that impact the efficiency, sustainability, and profitability of agricultural supply chains. Addressing these challenges is crucial for reducing post-harvest losses, ensuring food security, and maximizing the economic value of harvested crops. Here are some common challenges in post-harvest management:

1. **Infrastructure Limitations:** Inadequate storage facilities, transportation networks, and handling infrastructure pose significant challenges to post-harvest management, particularly in rural and remote areas. Limited access to cold storage, refrigerated transportation, and packaging facilities results in increased spoilage, quality deterioration, and market losses.

2. **Technological Barriers:** Limited access to appropriate post-harvest technologies and equipment hinders the adoption of modern storage, processing, and preservation methods. Small-scale farmers and resource-constrained regions often lack access to affordable and suitable technologies, limiting their ability to preserve the quality and value of harvested crops.

3. **Lack of Awareness and Training:** Insufficient knowledge and awareness about best practices in post-harvest management among farmers, producers, and stakeholders contribute to high levels of losses and inefficiencies. Lack of training programs, extension services, and educational resources further exacerbate the problem, hindering the adoption of improved handling, storage, and processing techniques.

4. **Pest and Disease Management:** Pests, pathogens, and microbial contaminants pose significant risks to post-harvest quality and safety, leading to spoilage, contamination, and rejection of crops. Inadequate pest control measures, improper sanitation practices, and suboptimal hygiene standards contribute to increased post-harvest losses and food safety concerns.

5. **Market Access and Trade Barriers:** Limited market access, trade restrictions, and regulatory barriers impede the efficient flow of agricultural products from producers to consumers. Inefficient distribution networks, lack of market information, and price volatility hinder farmers' ability to sell their produce at fair prices, resulting in reduced income and profitability.

6. **Environmental Factors:** Environmental conditions such as temperature fluctuations, humidity levels, and extreme weather events can impact post-harvest handling and storage practices. High temperatures accelerate ripening and decay, while excessive humidity promotes mold growth and spoilage. Climate variability and natural disasters further exacerbate the vulnerability of post-harvest operations.

7. **Financial Constraints:** Limited access to credit, financing, and investment capital restricts farmers' ability to invest in modern post-harvest infrastructure, equipment, and technologies. High upfront costs associated with cold storage, processing facilities, and packaging materials pose barriers to adoption, particularly for smallholders and marginalized producers.

8. **Supply Chain Inefficiencies:** Fragmented supply chains, lack of coordination among stakeholders, and inefficient logistics contribute to delays, losses, and quality deterioration in post-harvest operations. Poor coordination between producers, processors, distributors, and retailers leads to bottlenecks, stock outs, and mismatches in supply and demand.

9. **Quality Control and Standards:** Inconsistent quality standards, certification requirements, and food safety regulations create challenges for producers in meeting market requirements and export criteria. Lack of compliance with quality control measures, hygiene standards, and traceability systems results in rejected shipments, product recalls, and reputational damage.

10. **Post-Harvest Losses and Waste:** Post-harvest losses and waste occur at various stages of the supply chain, from harvesting to consumption, due to inefficiencies, mishandling, and inadequate storage conditions. These losses represent a missed opportunity to feed the growing global population, exacerbate food insecurity, and contribute to environmental degradation.

Conclusion

In conclusion, effective post-harvest management is essential for ensuring food security, economic sustainability, and environmental conservation in agriculture. By implementing appropriate harvesting techniques, storage methods, processing technologies, and distribution strategies, stakeholders can minimize losses, maximize value, and optimize resource utilization throughout the food supply chain. Addressing challenges such as infrastructure constraints, technological barriers, and market dynamics requires coordinated efforts from governments, development agencies, research institutions, and private sector stakeholders. By investing in post-harvest infrastructure, promoting technological innovation, and fostering market-oriented policies, we can enhance the resilience and efficiency of agricultural systems, ultimately improving livelihoods and food security for all.

References

Atanda, S. A., Pessu, P. O., Agoda, S., Isong, I. U., & Ikotun, I. (2011). The concepts and problems of post–harvest food losses in perishable crops. African Journal of Food Science, 5(11), 603-613.

Babu, P. J., Saranya, S., Longchar, B., & Rajasekhar, A. (2022). Nanobiotechnology-mediated sustainable agriculture and post-harvest management. Current Research in Biotechnology, 4, 326-336.

Hengsdıjk, H., & De Boer, W. J. (2017). Post-harvest management and post-harvest losses of cereals in Ethiopia. Food Security, 9, 945-958.

Kiaya, V. (2014). Post-harvest losses and strategies to reduce them. Technical Paper on Postharvest Losses, Action Contre la Faim (ACF), 25(3), 1-25.

Kimatu, J. N., McConchie, R., Xie, X., & Nguluu, S. N. (2012). The significant role of post-harvest management in farm management, aflatoxin mitigation and food security in Sub-Saharan Africa. Greener Journal of Agricultural Sciences, 2(6), 279-288.

Kumar, V., & Iqbal, N. (2020). Post-harvest pathogens and disease management of horticultural crop: A brief review. Plant Arch, 20, 2054-2058.

21

Post-Harvest Management of Horticultural Crops

Laxmi P S[1], Mirudhula Sekar[2], Anushka Singh[3] and Durgesh Kumar[4]

[1]*Department of Post harvest Management, Kerala Agriculture University Thrissur, Kerala*
[2]*Department of Spices and Plantation Crops, Tamil Nadu Agricultural University, Coimbatore, Tamil Nadu*
[3]*Department of Horticulture, K.P.H.E.I. Affiliated by Prof. Rajendra Singh (Rajju Bhaiya) University, Prayagraj, Uttar Pradesh*
[4]*Department of Agronomy, CSAUAT, Kanpur, Uttar Pradesh*

Abstract

Post-harvest management of horticultural crops is a critical aspect of agriculture aimed at preserving the quality, freshness, and nutritional value of fruits, vegetables, and ornamental plants from the time of harvest until consumption. This abstract highlights the importance of effective post-harvest management practices in minimizing losses, ensuring food security, and maximizing economic value in the horticultural sector. It emphasizes the diverse range of challenges faced, including perishability, susceptibility to physical damage, and vulnerability to pests and diseases. The abstract also underscores the significance of infrastructure development, technological innovation, and capacity building in addressing these challenges and improving post-harvest handling, storage, and distribution systems. Overall, the abstract provides a concise overview of the importance and complexities of post-harvest management in horticulture, laying the foundation for further research and action in this critical area of agriculture.

Keywords: *storage, pests, damage, economics*

Introduction

Horticultural crops, encompassing fruits, vegetables, flowers, and ornamental plants, play a significant role in global agriculture, economy, and nutrition. However, their perishable nature poses challenges in maintaining quality and extending shelf life after harvest. Post-harvest management practices are crucial to minimize losses, ensure food safety, and meet consumer demands. This article delves into the importance of post-harvest management in horticulture, highlighting key strategies, technologies, and innovations aimed at preserving quality and enhancing market value.

Importance of Post-Harvest Management

Post-harvest management plays a critical role in the overall agricultural supply chain, particularly for horticultural crops, due to their perishable nature and high susceptibility to post-harvest losses. Understanding the importance of post-harvest management is essential for ensuring food security, reducing waste, and maximizing economic returns. Here are some key points highlighting its significance

1. **Minimizing Losses:** Post-harvest losses account for a significant portion of global food wastage. In developing countries, inadequate post-harvest management practices result in substantial losses of horticultural crops, depriving millions of people of nutritious food and undermining efforts to alleviate hunger and malnutrition. Effective post-harvest management helps minimize losses by preserving the quality and extending the shelf life of perishable produce.

2. **Ensuring Food Safety:** Improper handling, storage, and transportation of horticultural crops can lead to contamination by pathogens, pesticides, and toxins, posing serious health risks to consumers. By implementing food safety measures such as cleaning, sanitization, and temperature control, post-harvest management systems mitigate these risks and ensure the delivery of safe and wholesome food to the market.

3. **Meeting Market Demands:** Consumer preferences for fresh, high-quality produce are driving the demand for efficient post-harvest management practices. Timely harvesting, proper handling, and effective packaging techniques help maintain the freshness, flavour, and nutritional value of horticultural crops, meeting market demands and enhancing consumer satisfaction. Moreover, improved post-harvest management contributes to the development of value-added products and expands market opportunities for farmers.

4. **Enhancing Economic Returns:** Post-harvest losses not only impact food security but also result in significant economic losses for farmers, traders, and other stakeholders along the supply chain. By minimizing losses and improving product quality, post-harvest management practices enhance the economic viability of horticultural farming operations. Additionally, value-added activities such as processing, packaging, and marketing of horticultural produce generate employment opportunities and contribute to rural livelihoods.
5. **Supporting Sustainable Agriculture:** Sustainable post-harvest management practices promote resource efficiency, reduce environmental impact, and support the transition towards more resilient and environmentally friendly agricultural systems. By adopting technologies and strategies that minimize waste, conserve energy, and reduce greenhouse gas emissions, stakeholders can contribute to sustainable development goals and mitigate the adverse effects of climate change on agriculture.

Key Post-Harvest Management Practices

Effective post-harvest management practices are essential for maintaining the quality, safety, and marketability of horticultural crops. These practices encompass a range of activities from harvesting to storage and transportation. Here are some key post-harvest management practices

1. **Harvesting at the Right Stage:** Harvesting horticultural crops at the appropriate stage of maturity is crucial for ensuring optimal quality and shelf life. Different crops have specific indicators of maturity such as colour, size, and firmness. Harvesting too early or too late can result in poor quality and reduced storability. Training farmers to recognize these indicators and harvest at the right time is essential.
2. **Gentle Handling:** Minimizing physical damage during harvesting and handling is vital to prevent bruising, cuts, and other injuries that can accelerate spoilage and reduce market value. Using sharp, clean tools and handling produce carefully to avoid mechanical damage is important. Additionally, providing proper training to workers involved in harvesting and handling operations can help minimize damage.
3. **Temperature Management:** Maintaining the optimal temperature is critical for preserving the quality and extending the shelf life of horticultural crops. Rapid cooling immediately after harvest helps to slow down metabolic processes and inhibit microbial growth. Cold

storage facilities, refrigerated transportation, and pre-cooling techniques such as hydro cooling and forced-air cooling are employed to maintain optimal temperatures throughout the supply chain.

4. **Cleaning and Sanitation:** Thorough cleaning and sanitation of harvested produce help remove dirt, debris, and surface contaminants, reducing microbial load and minimizing the risk of post-harvest diseases. Washing, brushing, and immersion in sanitizing solutions are common methods used to maintain hygiene and food safety. Proper sanitation of equipment, containers, and storage facilities is also essential to prevent cross-contamination.

5. **Packaging:** Choosing appropriate packaging materials and techniques is crucial for protecting horticultural crops during storage and transportation. Packaging should provide adequate ventilation to prevent moisture build-up while protecting the produce from physical damage and environmental factors. Modified atmosphere packaging (MAP) and controlled atmosphere storage (CAS) are advanced packaging methods used to extend shelf life by controlling gas composition and humidity levels.

6. **Quality Grading and Sorting:** Sorting and grading horticultural crops based on size, colour, shape, and quality parameters help standardize the product and meet market requirements. Automated sorting systems equipped with optical sensors and computer vision technology enable fast and accurate grading, reducing labour costs and improving efficiency.

7. **Post-Harvest Treatments:** Various post-harvest treatments such as hot water treatment, fungicide application, and irradiation are employed to control post-harvest diseases, enhance microbial safety, and extend shelf life. These treatments are applied based on the specific requirements of different crops and the prevailing market regulations.

8. **Market Access and Distribution:** Efficient distribution channels and market access are essential for delivering horticultural produce to consumers in a timely manner. Establishing cold chain logistics, developing market linkages, and collaborating with retailers and wholesalers help ensure that the produce reaches its destination fresh and in optimal condition.

Technological Innovations in Post-Harvest Management

Technological innovations have revolutionized post-harvest management practices in horticulture, offering solutions to improve efficiency, reduce losses, and enhance product quality. Here are some notable innovations in post-harvest management

1. **Smart Packaging Solutions:** Advanced packaging materials embedded with sensors, indicators, and antimicrobial agents offer real-time monitoring of product quality, freshness, and safety. These smart packaging technologies provide valuable information on storage conditions, shelf life predictions, and potential spoilage, enabling timely interventions to maintain product integrity and reduce waste.
2. **Cold Chain Management Systems:** Integrated cold chain management systems utilize IoT (Internet of Things) devices, RFID (Radio Frequency Identification) tags, and data analytics to monitor temperature, humidity, and location of horticultural produce throughout the supply chain. By ensuring compliance with temperature requirements, reducing temperature fluctuations, and minimizing losses due to spoilage, these systems improve the efficiency and reliability of cold chain logistics.
3. **Precision Agriculture Techniques:** Precision agriculture technologies such as remote sensing, GIS (Geographic Information Systems), and drones enable accurate monitoring of crop health, maturity, and yield prediction. By providing real-time data on field conditions, pest infestations, and crop growth stages, precision agriculture enhances decision-making in post-harvest operations, optimizing harvesting schedules, and minimizing losses.
4. **Biological Control Agents:** Biological control agents such as antagonistic microorganisms, biofungicides, and biopesticides offer environmentally friendly alternatives to synthetic chemicals for managing post-harvest diseases and pests. These bio-based interventions suppress pathogens, reduce chemical residues, and contribute to sustainable post-harvest management practices, ensuring food safety and environmental protection.
5. **Blockchain Technology:** Blockchain technology provides a secure and transparent platform for recording and tracking the entire supply chain journey of horticultural produce, from farm to fork. By enabling traceability, authentication, and verification of product information, blockchain enhances transparency, reduces food fraud, and builds consumer trust in the quality and safety of horticultural products.
6. **Advanced Storage Technologies:** Innovations in storage technologies such as controlled atmosphere storage (CAS), dynamic controlled

atmosphere (DCA), and vacuum cooling systems help extend the shelf life of horticultural crops by regulating oxygen, carbon dioxide, and humidity levels inside storage chambers. These technologies slow down respiration rates, inhibit microbial growth, and maintain product quality, ensuring that the produce remains fresh and marketable for longer periods.

7. **Robotics and Automation:** Robotic systems equipped with computer vision, machine learning, and robotic arms are increasingly being used for sorting, grading, and packaging of horticultural produce. These automated systems offer higher accuracy, speed, and efficiency compared to manual labour, reducing labour costs and improving productivity in post-harvest operations.

8. **Energy-Efficient Solutions:** Energy-efficient post-harvest technologies such as solar-powered cold storage units, evaporative cooling systems, and energy recovery ventilators help reduce energy consumption and operating costs while maintaining optimal storage conditions for horticultural produce. These solutions promote sustainability and resilience in post-harvest management practices, particularly in off-grid or resource-constrained environments.

Challenges and Future Directions

Despite significant advancements, post-harvest management in horticulture continues to face various challenges that necessitate concerted efforts and innovative solutions. Addressing these challenges and charting future directions is essential for ensuring food security, reducing waste, and enhancing sustainability. Here are some key challenges and potential future directions:

Challenges

1. **Infrastructure Constraints:** Inadequate infrastructure, particularly in rural areas and developing countries, hampers efficient post-harvest handling, storage, and transportation of horticultural crops. Limited access to cold storage facilities, reliable transportation networks, and market infrastructure undermines efforts to minimize losses and meet market demands.

2. **Knowledge and Skills Gap:** A lack of awareness, training, and technical expertise among farmers, extension workers, and other stakeholders poses challenges in adopting best practices and innovative technologies in post-harvest management. Bridging the knowledge gap and providing capacity-building initiatives are essential for promoting the adoption of effective post-harvest practices.

3. **Market Access Barriers:** Smallholder farmers often face challenges in accessing markets due to limited market information, poor market linkages, and inadequate bargaining power. Addressing market access barriers and empowering farmers with market intelligence, market infrastructure, and value chain partnerships can help improve their competitiveness and profitability.

4. **Food Safety Concerns:** Ensuring food safety throughout the post-harvest handling and distribution process remains a critical challenge. Contamination by pathogens, pesticides, and toxins poses risks to consumer health and undermines consumer confidence. Strengthening food safety regulations, promoting good agricultural practices, and investing in quality assurance systems are essential for mitigating food safety risks.

5. **Climate Change Impacts:** Climate change exacerbates existing challenges in post-harvest management by altering weather patterns, increasing temperature variability, and affecting crop yields and quality. Adapting post-harvest management practices to climate change impacts, investing in climate-resilient infrastructure, and promoting climate-smart agriculture approaches are necessary for building resilience in horticultural systems.

Future Directions

1. **Investment in Infrastructure:** Prioritizing investments in cold storage facilities, transportation networks, and market infrastructure is essential for improving post-harvest management efficiency and reducing losses. Public-private partnerships, international collaborations, and innovative financing mechanisms can mobilize resources for infrastructure development in horticulture.

2. **Capacity Building and Training:** Strengthening extension services, farmer training programs, and vocational education initiatives is crucial for enhancing the knowledge and skills of stakeholders in post-harvest management. Training on best practices, technological innovations, and market-oriented approaches can empower farmers and enhance their resilience to challenges.

3. **Market Diversification:** Diversifying market options and exploring new market channels such as e-commerce platforms, agri-tech start-ups, and direct-to-consumer sales can expand market access for smallholder farmers and improve their income opportunities. Leveraging digital

technologies and market intelligence tools can facilitate market linkages and value chain integration.

4. **Adoption of Digital Solutions:** Embracing digital solutions such as mobile applications, remote sensing technologies, and blockchain-based traceability systems can enhance transparency, traceability, and efficiency in post-harvest management. Digital platforms enable real-time monitoring, data-driven decision-making, and value chain optimization, facilitating the transition towards more resilient and sustainable horticultural systems.

5. **Promotion of Circular Economy:** Adopting circular economy principles in post-harvest management can contribute to waste reduction, resource efficiency, and environmental sustainability. Implementing practices such as composting, recycling of packaging materials, and valorisation of by-products can minimize waste generation and promote resource conservation in horticultural value chains.

Conclusion

Efficient post-harvest management is indispensable for maintaining quality, safety, and marketability of horticultural crops. By implementing best practices, leveraging technological innovations, and addressing existing challenges, stakeholders can enhance the efficiency and sustainability of post-harvest operations, thereby contributing to food security, economic growth, and environmental conservation. Continuous research, capacity building, and policy support are vital for realizing the full potential of post-harvest management in horticulture and ensuring the availability of fresh, nutritious produce for present and future generations.

References

Devi, E.P., and Kumari, B. (2015). A review on prospects of pre-harvest application of bioagents in managing post-harvest diseases of horticultural crops. International Journal of Agriculture, Environment and Biotechnology, 8(4), 933.

Dhall, R. K. (2013). Ethylene in post-harvest quality management of horticultural crops: A review. Research & Reviews: A Journal of Crop Science and Technology, 2(2), 9-24.

Kader, A. A. (2002). Postharvest technology of horticultural crops (Vol. 3311). University of California Agriculture and Natural Resources.

Kader, A. A. and Rolle, R. S. (2004). The role of post-harvest management in assuring the quality and safety of horticultural produce (Vol. 152). Food & Agriculture Org..

Kumar, V. and Iqbal, N. (2020). Post-harvest pathogens and disease management of horticultural crop: A brief review. Plant Arch, 20, 2054-2058.

Lurie, S. (2010). Post-harvest heat treatments of horticultural crops. Horticultural Reviews (Am Soc Hortic Sci), 22(22), 91-121.

Meena, M. S., Kumar, A., Singh, K. M. and Meena, H. R. (2016). Farmers' attitude towards post-harvest issues of horticultural crops. Indian Research Journal of Extension Education, 9(3), 15-19.

Nath, A., Meena, L.R., Kumar, V., and Panwar, A. S. (2018). Postharvest management of horticultural crops for doubling farmer's income.

22

Organic Farming and Soil Health

Rajni Yadav[1], Anil Kumar[2], Sekhar Kumar[3] and Shwetank Shukla[4]

[1,2,3]Department of Soil Science, Chaudhary Charan Singh Haryana Agricultural University, Hisar, Haryana
[4]Department of Soil Science and Agricultural Chemistry, Acharya Narendra Deva University Agriculture and Technology, Kumarganj, Ayodhya

Abstract

Organic farming is increasingly recognized as a sustainable agricultural practice that prioritizes soil health, ecological balance, and long-term productivity. This article explores the intricate relationship between organic farming and soil health, detailing the principles, practices, benefits, challenges, and future prospects. Key organic farming practices such as crop rotation, cover cropping, composting, green manuring, and reduced tillage play crucial roles in enhancing soil structure, fertility, and biodiversity. While organic farming offers numerous benefits, including improved soil fertility, structure, microbial activity, and water management, it also faces challenges like nutrient management, pest control, soil erosion, and weed management. Through case studies and future perspectives, this article underscores the potential of organic farming to contribute to sustainable agriculture and environmental protection, advocating for continued research, policy support, and education to maximize its benefits.

Keywords: *organic, benefits, policy, nutrient, cover cropping*

Introduction

Organic farming has gained significant attention as a sustainable agricultural practice that emphasizes the health of the soil, ecosystems, and people. Unlike conventional farming, which relies heavily on synthetic chemicals and fertilizers, organic farming employs natural processes and inputs to maintain and improve soil health. This comprehensive article delves into the intricate relationship between organic farming and soil health, exploring the principles, practices, benefits, challenges, and future prospects.

Principles of Organic Farming

Organic farming is founded on a set of principles that aim to create a sustainable and harmonious agricultural system. These principles prioritize ecological balance, biodiversity, and the health of both the soil and the broader environment. Here are the core principles of organic farming

1. **Health**: The principle of health emphasizes maintaining and enhancing the health of the soil, plants, animals, humans, and the planet as a whole. Organic farming seeks to produce food that is nutritious and free from harmful chemicals, thereby promoting overall well-being.
2. **Ecology**: Organic farming practices are designed to work with natural ecological systems and cycles. This principle involves maintaining natural ecosystems and using sustainable methods that do not disrupt ecological balance. Techniques such as crop rotation, cover cropping, and biological pest control are employed to mimic natural processes and support the environment.
3. **Fairness**: Fairness in organic farming pertains to the equity and justice of the farming practices for all stakeholders, including farmers, workers, consumers, and the environment. This principle advocates for fair trade, proper working conditions, and the equitable distribution of resources and benefits.
4. **Care**: The principle of care focuses on responsible management of agricultural systems to protect the health and well-being of current and future generations. It involves taking a precautionary approach to avoid potential risks from agricultural practices and technologies, ensuring sustainability, and fostering resilience.
5. **Biodiversity**: Organic farming emphasizes the importance of maintaining and enhancing biodiversity. This includes preserving genetic diversity in crops and livestock, fostering diverse ecosystems, and supporting a wide range of species and natural habitats. Biodiversity is crucial for resilience against pests, diseases, and changing environmental conditions.
6. **Sustainability**: Sustainability is a cornerstone of organic farming, aiming for long-term agricultural productivity without depleting resources or causing environmental harm. Organic practices are designed to maintain soil fertility, reduce pollution, and use resources efficiently to ensure that farming can continue for future generations.

Practices in Organic Farming

Organic farming employs a variety of practices designed to maintain and improve soil health, promote biodiversity, and enhance ecological balance. These practices are essential to achieving the principles of organic farming and ensuring sustainable agricultural systems. Here are some key practices in organic farming:

1. Crop Rotation

- **Definition**: The practice of growing different types of crops in the same area across different seasons or years.
- **Benefits**: Breaks pest and disease cycles, improves soil structure and fertility, and reduces soil erosion. Different crops have varying nutrient requirements and root structures, which helps balance soil nutrient levels and promotes biodiversity.

2. Cover Cropping

- **Definition**: Planting cover crops, such as clover, rye, or legumes, during off-seasons when main crops are not grown.
- **Benefits**: Prevents soil erosion, enhances soil fertility through nitrogen fixation (particularly with leguminous cover crops), and improves soil organic matter. Cover crops also suppress weeds and provide habitat for beneficial insects.

3. Composting

- **Definition**: The process of recycling organic waste materials, such as crop residues, kitchen scraps, and animal manure, into nutrient-rich compost.
- **Benefits**: Adds organic matter to the soil, improves soil structure and water retention, and provides a slow-release source of nutrients. Composting also helps reduce the need for synthetic fertilizers and minimizes waste.

4. Green Manuring

- **Definition**: Growing specific plants (green manures) that are ploughed back into the soil to decompose and enrich it.
- **Benefits**: Increases soil organic matter, enhances nutrient availability, and improves soil structure. Green manures can also suppress weeds and prevent soil erosion.

5. Reduced Tillage

- **Definition**: Minimizing soil disturbance by reducing the frequency and intensity of tillage.
- **Benefits**: Preserves soil structure, reduces erosion, and maintains soil moisture. Reduced tillage also supports beneficial soil organisms and reduces the carbon footprint of farming operations.

6. Biological Pest Control

- **Definition**: Using natural predators, parasites, or pathogens to control pest populations.
- **Benefits**: Reduces the need for chemical pesticides, promotes a balanced ecosystem, and enhances biodiversity. Examples include introducing ladybugs to control aphids or using Bacillus thuringiensis (Bt) to target specific insect larvae.

7. Organic Mulching

- **Definition**: Applying organic materials, such as straw, leaves, or wood chips, to the soil surface.
- **Benefits**: Conserves soil moisture, suppresses weeds, regulates soil temperature, and adds organic matter as the mulch decomposes. Mulching also protects soil from erosion and compaction.

8. Integrated Livestock Management

- **Definition**: Integrating livestock into the farming system in a way that benefits both the animals and the crops.
- **Benefits**: Provides natural fertilizer through manure, enhances soil fertility, and promotes nutrient cycling. Livestock can also help with weed control and contribute to farm biodiversity.

9. Polyculture and Agroforestry

- **Definition**: Growing multiple crop species together (polyculture) or integrating trees and shrubs with crops and/or livestock (agroforestry).
- **Benefits**: Increases biodiversity, improves soil structure, enhances nutrient cycling, and provides habitat for beneficial organisms. These systems also help in pest management and can increase overall farm resilience.

10. Water Conservation Techniques

- **Definition**: Implementing methods to use water efficiently and sustainably.
- **Benefits**: Reduces water usage, maintains soil moisture, and prevents waterlogging and soil erosion. Techniques include drip irrigation, rainwater harvesting, and the use of drought-resistant crop varieties.

Soil Health in Organic Farming

Soil health is a cornerstone of organic farming, reflecting the condition of soil as a living, dynamic ecosystem. Healthy soil supports plant growth, maintains water and nutrient cycling, and enhances the resilience of the agricultural system. Organic farming practices are specifically designed to promote and sustain soil health through various means.

Key Indicators of Soil Health

1. Soil Organic Matter (SOM)

- **Importance**: SOM is crucial for maintaining soil structure, water retention, and nutrient availability. It comprises decomposed plant and animal residues, which form humus.
- **Organic Practices**: Organic farming increases SOM through composting, green manuring, and cover cropping. These practices add organic residues to the soil, which decompose and contribute to humus formation.

2. Soil Structure

- **Importance**: Good soil structure improves root growth, water infiltration, and air exchange. It involves the arrangement of soil particles into aggregates.
- **Organic Practices**: Reduced tillage, organic amendments, and crop rotations enhance soil aggregation and porosity. Maintaining a stable soil structure reduces erosion and compaction.

3. Soil Fertility

- **Importance**: Fertile soil supplies essential nutrients in adequate amounts and balances to support plant growth. Nutrient availability and cycling are critical for maintaining soil fertility.
- **Organic Practices**: Organic farms rely on natural nutrient sources such as compost, animal manure, and green manures. These inputs provide a slow-release source of nutrients, enhancing soil fertility over time.

4. Microbial Activity

- **Importance**: Soil microbes play a vital role in nutrient cycling, organic matter decomposition, and disease suppression. A diverse and active microbial community is indicative of healthy soil.
- **Organic Practices**: Organic farming promotes microbial diversity and activity by avoiding synthetic chemicals and using organic inputs. Practices like composting and reduced tillage create a conducive environment for beneficial microbes.

5. Erosion Control

- **Importance**: Preventing soil erosion is essential to maintain topsoil and soil fertility. Erosion can lead to the loss of nutrients and organic matter, degrading soil health.
- **Organic Practices**: Cover cropping, maintaining ground cover, and reduced tillage help prevent soil erosion. These practices protect the soil surface from wind and water erosion.

Benefits of Organic Farming to Soil Health

Organic farming provides numerous benefits to soil health, enhancing its structure, fertility, biodiversity, and overall resilience. These benefits are achieved through sustainable practices that prioritize ecological balance and the natural nutrient cycling of the soil ecosystem. Here are the key benefits of organic farming to soil health:

1. Enhanced Soil Fertility

Mechanism: Organic farming practices such as the use of compost, green manure, and animal manure contribute significantly to soil fertility. These organic inputs decompose slowly, releasing nutrients gradually and providing a sustained supply of essential elements like nitrogen, phosphorus, and potassium.

Outcome: This gradual nutrient release improves nutrient availability and uptake by plants, leading to healthier crops. Enhanced soil fertility also reduces the dependency on synthetic fertilizers, fostering a more sustainable agricultural system.

2. Improved Soil Structure

Mechanism: Practices like reduced tillage, composting, and cover cropping promote the formation of stable soil aggregates. These aggregates enhance soil porosity, allowing better root penetration and facilitating air and water movement within the soil.

Outcome: Improved soil structure reduces soil compaction, enhances water infiltration and retention, and minimizes runoff and erosion. This leads to more robust plant growth and higher resistance to drought conditions.

3. Increased Soil Organic Matter (SOM)

Mechanism: The incorporation of organic materials such as crop residues, cover crops, and compost increases the organic matter content of the soil. Organic matter acts as a reservoir of nutrients and improves the soil's physical properties.

Outcome: Higher levels of SOM improve soil fertility, water-holding capacity, and microbial activity. This creates a healthier soil ecosystem that supports plant growth and resists environmental stresses.

4. Enhanced Microbial Activity and Biodiversity

Mechanism: Organic farming avoids synthetic chemicals, which can be harmful to soil microbes. Instead, it fosters microbial diversity through organic inputs and minimal soil disturbance.

Outcome: A diverse and active microbial community enhances nutrient cycling, organic matter decomposition, and disease suppression. Soil biodiversity, including beneficial insects and earthworms, contributes to the overall health and resilience of the soil ecosystem.

5. Improved Water Management

Mechanism: Organic practices such as mulching, cover cropping, and maintaining soil organic matter improve the soil's ability to retain moisture and enhance water infiltration.

Outcome: Improved water management reduces the need for irrigation, lowers the risk of waterlogging and erosion, and enhances plant water availability. This is particularly beneficial in areas prone to drought or irregular rainfall.

6. Carbon Sequestration

Mechanism: Organic farming practices increase the amount of organic carbon stored in the soil. Techniques like reduced tillage, cover cropping, and the use of organic amendments promote the sequestration of carbon from the atmosphere into the soil.

Outcome: Increased carbon sequestration helps mitigate climate change by reducing greenhouse gas levels. It also improves soil structure and fertility, contributing to long-term soil health.

7. Erosion Control

Mechanism: Practices such as cover cropping, maintaining ground cover, and reduced tillage help protect the soil surface from erosion by wind and water.

Outcome: Effective erosion control preserves topsoil, maintains soil fertility, and prevents the loss of valuable soil nutrients. This leads to more sustainable land use and agricultural productivity.

8. Reduction of Soil Contamination

Mechanism: Organic farming eliminates the use of synthetic pesticides and fertilizers, which can contaminate soil and water resources.

Outcome: Reducing soil contamination improves soil health and safety for crop production, contributing to healthier ecosystems and food supplies.

Challenges in Organic Farming

Despite its numerous benefits, organic farming faces several significant challenges. These challenges can impact productivity, profitability, and the widespread adoption of organic practices. Understanding and addressing these challenges is crucial for the growth and success of organic agriculture. Here are some of the key challenges in organic farming:

1. Nutrient Management

Challenge: Organic farming relies on natural nutrient sources, such as compost, animal manure, and green manures, which can be less predictable and slower-acting than synthetic fertilizers. Balancing nutrient supply with crop demand can be difficult, particularly in high-yield systems.

Solution: Effective nutrient management plans, including regular soil testing, crop rotations, and diversified organic amendments, can help optimize nutrient availability. Research into more efficient organic fertilizers and nutrient management techniques is also essential.

2. Pest and Disease Control

Challenge: Without synthetic pesticides, organic farmers must rely on integrated pest management (IPM) strategies, which can be more labor-intensive and require extensive knowledge. Organic pest control methods, such as biological control, crop rotations, and the use of resistant varieties, may not always provide immediate results.

Solution: Education and training in IPM techniques, combined with ongoing research into organic pest control solutions, can help farmers manage pests and diseases effectively. Developing and promoting natural biopesticides and beneficial insect habitats can also enhance pest control.

3. Soil Erosion

Challenge: While organic practices generally reduce erosion, certain conditions, such as heavy rainfall, steep slopes, and bare soil during certain crop rotations, can still pose significant risks. Erosion can lead to the loss of topsoil and nutrients, degrading soil health.

Solution: Implementing erosion control measures, such as maintaining continuous ground cover, using contour farming, and applying organic mulches, can mitigate these risks. Cover cropping and reduced tillage practices are particularly effective in preventing soil erosion.

4. Weed Management

Challenge: Organic farms often struggle with weed control, as they do not use herbicides. Weeds can compete with crops for nutrients, water, and light, reducing yields. Mechanical weeding, mulching, and cover cropping are labor-intensive and may not be entirely effective.

Solution: Developing more efficient mechanical weeding tools and techniques, along with research into natural weed suppressants and competitive crop varieties, can help manage weeds. Integrating livestock for grazing and using cover crops can also reduce weed pressure.

5. Yield Variability

Challenge: Organic farming systems can experience greater yield variability compared to conventional systems due to factors like nutrient availability, pest and disease pressure, and weather conditions. This variability can impact profitability and farmer confidence.

Solution: Improving organic farming practices through research and innovation can help stabilize yields. Diversifying crops and implementing robust risk management strategies can also mitigate the impact of yield variability.

6. Labor Intensity

Challenge: Organic farming often requires more manual labor for tasks such as weeding, composting, and pest control. This increased labor requirement can lead to higher production costs and challenges in scaling up organic operations.

Solution: Mechanization and technological advancements tailored to organic systems can reduce labor intensity. Training programs and cooperative labor arrangements can also help manage labor demands.

7. Economic Viability

Challenge: Organic products often command higher prices, but the cost of production can also be higher due to labor, certification, and input costs. Transitioning to organic farming involves a period where yields may decline, further impacting economic viability.

Solution: Government subsidies, financial incentives, and market development initiatives can support farmers during the transition period. Increasing consumer awareness and demand for organic products can also enhance market opportunities and profitability.

8. Certification and Regulation

Challenge: Obtaining organic certification involves stringent standards, thorough documentation, and regular inspections, which can be time-consuming and costly for farmers. Navigating different certification requirements in various regions can also be complex.

Solution: Simplifying the certification process and providing support and resources for farmers can ease the burden. Harmonizing standards internationally and creating more accessible certification schemes can also help.

9. Market Access

Challenge: Organic farmers may face challenges in accessing markets, particularly if they are located in remote areas or lack the infrastructure to reach urban consumers. Competing with conventional products and securing fair prices can also be difficult.

Solution: Developing local and regional markets, enhancing supply chain infrastructure, and promoting direct-to-consumer sales channels such as farmers' markets and community-supported agriculture (CSA) programs can improve market access. Marketing initiatives that highlight the benefits of organic products can also attract consumers.

Future Prospects

The future of organic farming and soil health looks promising, driven by increasing consumer demand for sustainable and healthy food, advances in organic farming techniques, and growing recognition of the environmental and health benefits of organic systems. Key areas of focus for future development include:

1. **Research and Innovation**: Continued research into organic farming practices and soil health will provide new insights and technologies to improve efficiency and productivity.
2. **Policy Support**: Government policies that support organic farming through subsidies, research funding, and market development can encourage more farmers to adopt organic practices.
3. **Education and Training**: Expanding education and training programs for farmers on organic practices and soil health management will enhance the adoption of sustainable techniques.
4. **Market Development**: Developing markets for organic products, including certification and consumer awareness campaigns, will drive demand and support organic farmers.
5. **Climate Change Adaptation**: Organic farming can play a critical role in climate change adaptation through improved soil health, carbon sequestration, and resilience to extreme weather events.

Conclusion

Organic farming presents a sustainable approach to agriculture that prioritizes soil health, ecological balance, and long-term productivity. By employing practices such as crop rotation, composting, and reduced tillage, organic farming enhances soil structure, fertility, and biodiversity. Despite challenges like nutrient management and pest control, the benefits to soil health and the environment are substantial. As research and innovation continue to advance, organic farming holds significant promise for a sustainable and resilient agricultural future.

References

Aulakh, C. S., Sharma, S., Thakur, M., & Kaur, P. (2022). A review of the influences of organic farming on soil quality, crop productivity and produce quality. Journal of Plant Nutrition, 45(12), 1884-1905.

Biswas, S., Ali, M. N., Goswami, R., & Chakraborty, S. (2014). Soil health sustainability and organic farming: A review. Journal of Food Agriculture and Environment, 12(3-4), 237-243.

M. Tahat, M., M. Alananbeh, K., A. Othman, Y., & I. Leskovar, D. (2020). Soil health and sustainable agriculture. Sustainability, 12(12), 4859.

Montgomery, D. R., & Biklé, A. (2021). Soil health and nutrient density: beyond organic vs. conventional farming. Frontiers in Sustainable Food Systems, 5, 417.

Reeve, J. R., Hoagland, L. A., Villalba, J. J., Carr, P. M., Atucha, A., Cambardella, C., ... & Delate, K. (2016). Organic farming, soil health, and food quality: considering possible links. Advances in agronomy, 137, 319-367.

23

Soil-Plant-Water Relationship

Rajni Yadav[1], Anil Kumar[2], Dharam Pal[3] and Shwetank Shukla[4]

[1,2,3]Department of Soil Science, Chaudhary Charan Singh Haryana Agricultural University, Hisar, Haryana
[4]Department of Soil Science and Agricultural Chemistry, Acharya Narendra Deva university Agriculture and Technology, Kumarganj, Ayodhya Uttar Pradesh

Abstract

The soil-plant-water relationship is a critical aspect of agricultural science, encompassing the interactions between soil properties, plant health, and water dynamics. This article examines the intricate connections that influence plant growth and crop productivity, highlighting the roles of soil texture, structure, and organic matter in water retention and availability. It explores how plants uptake water through root systems and how transpiration affects water distribution within the soil. The study also addresses the impact of irrigation practices and soil management techniques on optimizing water use efficiency in agricultural systems. Understanding these relationships is essential for developing sustainable farming practices that enhance crop yield while conserving water resources.

Keywords: *soil, plant, crop, productivity, water*

The soil-plant-water relationship is fundamental to agricultural productivity and environmental sustainability. Understanding this relationship involves examining the interactions between soil properties, plant physiological processes, and water dynamics. This knowledge is crucial for optimizing crop yields, improving soil health, and managing water resources efficiently. This article provides an in-depth analysis of the soil-plant-water relationship, covering key aspects such as soil properties, plant water uptake, transpiration, irrigation practices, and sustainable soil and water management techniques.

Soil Properties and Water Dynamics

Understanding soil properties and water dynamics is essential for optimizing agricultural productivity and sustainability. Soil's physical and chemical characteristics profoundly influence how water is retained, transmitted, and utilized by plants. This section delves into the various soil properties that affect water dynamics and explores the mechanisms through which water interacts with soil.

1. Soil Texture and Structure

Soil Texture

- **Definition**: Soil texture refers to the proportion of sand, silt, and clay particles in the soil. This composition affects the soil's ability to retain and transmit water.
- **Sand**: Coarse particles with a diameter between 0.05 and 2.0 mm. Sand has large pores that allow rapid water infiltration but low water retention.
- **Silt**: Medium-sized particles with a diameter between 0.002 and 0.05 mm. Silt retains more water than sand and has moderate permeability.
- **Clay**: Fine particles with a diameter less than 0.002 mm. Clay has very small pores, leading to high water retention but low permeability.

Soil Structure

- **Definition**: Soil structure refers to the arrangement of soil particles into aggregates or clumps, which can affect water movement and retention.
- **Types of Soil Structure**:
  - **Granular**: Small, rounded aggregates typical in topsoil, enhancing water infiltration and root growth.
  - **Blocky**: Irregular, block-like structures common in subsoil, which can impede water movement and root penetration.
  - **Platy**: Thin, flat plates that can restrict water movement and root growth, often resulting from soil compaction.

2. Soil Water Content and Availability

Soil Moisture Content

- **Field Capacity**: The maximum amount of water soil can hold after excess water has drained away. It represents the soil's ability to retain water for plant use.

- **Permanent Wilting Point**: The moisture level at which plants can no longer extract water, leading to wilting and potential plant death.
- **Available Water Capacity**: The range of water content between field capacity and permanent wilting point, indicating the water available for plant uptake.

Soil Water Potential

- **Definition**: The energy required for plants to extract water from the soil. It includes various components such as matric potential, osmotic potential, and gravitational potential.
- **Matric Potential**: The force exerted by soil particles on water, affecting water availability to plants. It is influenced by soil texture and structure.
- **Osmotic Potential**: The effect of dissolved salts and other solutes on water availability, particularly important in saline soils.
- **Gravitational Potential**: The influence of gravity on water movement, primarily affecting drainage and deep percolation.

3. Soil Water Retention and Movement

Soil Water Retention

- **Adhesion**: The attraction of water molecules to soil particles, which helps retain water in the soil.
- **Cohesion**: The attraction of water molecules to each other, aiding in the formation of continuous water films within the soil.
- **Capillary Action**: The movement of water through small pores in the soil, driven by the adhesive and cohesive forces. Capillary action is crucial for water movement in finer-textured soils like silt and clay.

Soil Water Movement

- **Infiltration**: The process by which water enters the soil surface. Soil texture, structure, and organic matter content significantly influence infiltration rates.
- **Percolation**: The downward movement of water through the soil profile. Percolation rates vary with soil texture, structure, and the presence of compacted layers.
- **Hydraulic Conductivity**: The ease with which water moves through soil pores. High hydraulic conductivity indicates rapid water movement,

typical in sandy soils, while low hydraulic conductivity is common in clayey soils.

4. Factors Affecting Soil Water Dynamics

Organic Matter

- **Role**: Organic matter improves soil structure, increases water holding capacity, and enhances nutrient availability. It contributes to the formation of stable soil aggregates and enhances microbial activity.
- **Impact**: Soils rich in organic matter retain more water and have better infiltration rates, reducing runoff and erosion.

Soil Compaction

- **Cause**: Compaction results from heavy machinery, livestock trampling, or repetitive tillage. It reduces pore space, impeding water infiltration and root growth.
- **Impact**: Compacted soils have lower infiltration rates, increased runoff, and reduced water availability to plants.

Soil Salinity

- **Cause**: Accumulation of salts in the soil, often due to improper irrigation practices or high evaporation rates in arid regions.
- **Impact**: High salinity reduces osmotic potential, making it difficult for plants to absorb water. It can lead to reduced crop yields and soil degradation.

Soil pH

- **Role**: Soil pH affects the solubility of nutrients and the activity of soil microorganisms. Extremes in pH can influence water retention and nutrient availability.
- **Impact**: Soils with very high or very low pH may have altered water dynamics and reduced microbial activity, affecting plant growth.

5. Interaction of Soil, Plant, and Water

Plant Root Systems

- **Types**: Different plants have varying root architectures, affecting their ability to access water. Taproots penetrate deeply, accessing water from lower soil layers, while fibrous roots spread out near the surface, making use of topsoil moisture.

- **Adaptations**: Plants adapt their root growth in response to soil moisture conditions. For example, roots may grow deeper in search of water during drought conditions.

Transpiration

- **Process**: Transpiration involves the movement of water from the soil through plant roots, stems, and leaves, eventually evaporating from leaf surfaces. It drives the uptake of water and nutrients from the soil.
- **Factors Influencing Transpiration**: Environmental factors such as light, temperature, humidity, and wind affect transpiration rates. Soil moisture availability also plays a critical role.

Soil-Water-Plant Interaction

- **Water Uptake**: Plants absorb water through root hairs, driven by the water potential gradient between the soil and root cells. Water moves through the plant via the xylem, reaching the leaves where it is used in photosynthesis and evaporates during transpiration.
- **Nutrient Transport**: Water in the soil dissolves nutrients, making them available for uptake by plant roots. Transpiration stream moves these nutrients to different parts of the plant.
- **Soil Health**: Healthy soil with good structure, sufficient organic matter, and balanced pH supports optimal water dynamics, enhancing plant growth and resilience to stress.

Soil Water Content and Availability

Soil water content and availability are critical factors that influence plant growth, agricultural productivity, and ecosystem health. Understanding these aspects of soil-water dynamics is essential for efficient water management in agriculture and environmental conservation. This section explores the concepts of soil water content, field capacity, wilting point, and available water capacity, highlighting their significance in soil-plant interactions and water resource management.

1. Soil Water Content

Definition: Soil water content refers to the amount of water present in the soil at a given time, expressed as a percentage of the soil's total weight or volume. It is a dynamic variable that fluctuates based on factors such as precipitation, evaporation, plant uptake, and soil characteristics.

Measurement: Soil water content can be measured using various techniques, including gravimetric methods, soil moisture sensors, and neutron probes. Gravimetric methods involve weighing soil samples before and after drying to determine water content.

Factors Affecting Soil Water Content

- **Soil Texture**: Fine-textured soils like clay hold more water than coarse-textured soils like sand due to their smaller pore spaces.
- **Soil Structure**: Well-aggregated soils with good structure retain more water than compacted soils with poor structure.
- **Climate**: Precipitation, temperature, humidity, and wind influence soil water content by affecting water input (precipitation) and output (evaporation and transpiration).
- **Vegetation**: Plants extract water from the soil through their roots, influencing soil water content. Dense vegetation can reduce soil water availability, especially during periods of high water demand.

2. Field Capacity

Definition: Field capacity is the maximum amount of water that soil can hold against gravity after excess water has drained away. It represents the soil's ability to retain water for plant use under saturated conditions.

Measurement: Field capacity is typically determined by saturating the soil with water and allowing it to drain until no more water is released, at which point the soil is considered to be at field capacity.

Significance: Field capacity indicates the upper limit of soil water availability for plant uptake. It provides a baseline for estimating irrigation requirements and understanding water holding capacity in different soil types.

3. Permanent Wilting Point

Definition: The permanent wilting point is the soil moisture level at which plants can no longer extract water, leading to irreversible wilting and potential plant death. It represents the lower limit of soil water availability for plant growth.

Measurement: The permanent wilting point is determined by allowing soil to dry out until plants show signs of wilting that do not recover even after watering.

Significance: The permanent wilting point indicates the critical threshold below which plants experience water stress and reduced growth. It helps determine irrigation scheduling and assess plant water needs.

4. Available Water Capacity

Definition: Available water capacity is the range of soil moisture content between field capacity and the permanent wilting point. It represents the amount of water that is readily available to plants for uptake and growth.

Calculation: Available water capacity is calculated as the difference between field capacity and the permanent wilting point. It is expressed in units such as inches or millimetres of water per foot or meter of soil depth.

Importance: Available water capacity is a crucial parameter for assessing soil fertility, crop suitability, and irrigation management. Soils with higher available water capacity can support more extensive root systems and sustain plant growth for longer periods between irrigation events.

5. Factors Influencing Soil Water Availability

Soil Texture: Sandy soils have lower water holding capacity but drain quickly, while clay soils have higher water holding capacity but slower drainage rates.

Soil Structure: Well-structured soils with good aggregation retain more water and support better root penetration than compacted soils.

Climate: Regions with high rainfall and humidity typically have higher soil moisture content, while arid regions may experience water deficits.

Land Use and Management: Vegetation cover, tillage practices, and irrigation methods can influence soil water availability and dynamics.

Root Distribution: The depth and density of plant roots affect water uptake and soil moisture distribution.

2. Plant Water Uptake and Transpiration

Root Systems and Water Uptake

Root Architecture: The arrangement of roots in the soil, which affects a plant's ability to absorb water and nutrients.

- **Taproot System**: A single main root with smaller lateral roots, common in dicots.
- **Fibrous Root System**: Numerous thin roots spreading out from the base of the plant, typical in monocots.

Root Hair Zone: The region where root hairs increase the surface area for water absorption. Root hairs absorb water through osmosis, driven by the water potential gradient between the soil and the root.

Transpiration Process

Transpiration: The process by which water is absorbed by plant roots, moves through the plant, and evaporates from the leaves. It plays a crucial role in nutrient transport, cooling the plant, and maintaining turgor pressure.

- **Stomata**: Small pores on leaf surfaces that regulate water loss and gas exchange. Stomatal opening and closing are influenced by environmental factors such as light, temperature, humidity, and CO_2 concentration.
- **Cohesion-Tension Theory**: Explains the movement of water through the plant. Water molecules stick together (cohesion) and to the walls of xylem vessels (adhesion), creating a continuous column of water from roots to leaves.

3. Water Management in Agriculture

Irrigation Practices

Surface Irrigation: Water is applied directly to the soil surface and allowed to infiltrate. Methods include furrow, basin, and border irrigation.

- **Advantages**: Low initial cost, suitable for a variety of crops.
- **Disadvantages**: Inefficient water use, potential for soil erosion and runoff.

Sprinkler Irrigation: Water is sprayed over the crops using a system of pipes and sprinklers.

- **Advantages**: More uniform water distribution, suitable for various terrains.
- **Disadvantages**: High initial cost, potential for water loss due to evaporation and wind drift.

Drip Irrigation: Water is delivered directly to the root zone through a network of tubes and emitters.

- **Advantages**: High water use efficiency, reduced evaporation and runoff, suitable for precise water management.
- **Disadvantages**: High initial cost, potential for clogging and maintenance issues.

Scheduling and Efficiency

Irrigation Scheduling: Determining the timing and amount of water to apply based on crop needs, soil moisture status, and weather conditions.

- **Soil Moisture Sensors**: Measure soil water content and provide data for irrigation decisions.
- **Weather-Based Systems**: Use weather data to estimate evapotranspiration rates and adjust irrigation accordingly.

Water Use Efficiency (WUE): The ratio of crop yield to the amount of water used. Improving WUE involves optimizing irrigation practices, selecting drought-tolerant crop varieties, and implementing soil conservation techniques.

4. Sustainable Soil and Water Management

Soil Conservation Practices

Cover Cropping: Planting cover crops, such as legumes or grasses, during off-seasons to protect and improve soil health.

- **Benefits**: Reduces soil erosion, enhances soil organic matter, improves water infiltration and retention.

Crop Rotation: Alternating different crops in a sequence to maintain soil fertility and reduce pest and disease build-up.

- **Benefits**: Enhances soil structure, diversifies root systems, breaks pest and disease cycles.

Reduced Tillage: Minimizing soil disturbance to maintain soil structure and organic matter.

- **Benefits**: Reduces erosion, conserves soil moisture and enhances microbial activity.

Integrated Water Management

Rainwater Harvesting: Collecting and storing rainwater for agricultural use.

- **Benefits**: Provides a supplementary water source, reduces dependency on groundwater, and enhances resilience to drought.

Soil Moisture Conservation: Techniques such as mulching and maintaining ground cover to reduce evaporation and improve soil moisture retention.

- **Benefits**: Enhances water availability, reduces irrigation needs and improves soil health.

Efficient Irrigation Systems: Implementing technologies such as drip irrigation and smart irrigation controllers to optimize water use.

- **Benefits**: Increases water use efficiency, reduces water waste, supports sustainable water management.

5. The Role of Soil-Plant-Water Relationship in Climate Change Adaptation

Impact of Climate Change

Temperature Variability: Changes in temperature affect evapotranspiration rates and soil moisture dynamics, impacting water availability for crops.

Altered Precipitation Patterns: Irregular rainfall can lead to water stress, affecting crop growth and yield.

Increased Frequency of Extreme Events: Droughts, floods, and heatwaves pose significant challenges to agricultural productivity and soil health.

Adaptive Strategies

Drought-Tolerant Crops: Developing and planting crop varieties that can withstand water stress and maintain productivity under drought conditions.

Soil Health Improvement: Enhancing soil organic matter and structure to improve water retention and resilience to climate impacts.

Water-Smart Agriculture: Implementing water-efficient practices and technologies to optimize water use and reduce vulnerability to water scarcity.

Climate-Resilient Infrastructure: Investing in infrastructure such as rainwater harvesting systems, efficient irrigation networks, and drainage systems to manage water resources effectively.

Conclusion

The soil-plant-water relationship is a complex and dynamic system that is crucial for sustainable agriculture. Understanding the interactions between soil properties, plant physiological processes, and water dynamics allows for the development of effective management practices that enhance soil health, optimize water use, and improve crop productivity. By adopting sustainable soil and water management techniques, farmers can mitigate the impacts of climate change, conserve natural resources, and ensure long-term agricultural sustainability. Continued research and innovation in this field are essential to address the challenges and harness the potential benefits of the soil-plant-water relationship.

References

Augé, R. M. (2004). Arbuscular mycorrhizae and soil/plant water relations. Canadian Journal of Soil Science, 84(4), 373-381.

Deng, Z., Guan, H., Hutson, J., Forster, M. A., Wang, Y., & Simmons, C. T. (2017). A vegetation-focused soil-plant-atmospheric continuum model to study hydrodynamic soil-plant water relations. Water Resources Research, 53(6), 4965-4983.

Kirkham, M. B. (2023). Principles of soil and plant water relations. Elsevier.

Mengel, K., Kirkby, E. A., Kosegarten, H., & Appel, T. (2001). Plant water relationships. Principles of plant nutrition, 181-242.

Wu, X., Shi, J., Zuo, Q., Zhang, M., Xue, X., Wang, L., & Ben-Gal, A. (2021). Parameterization of the water stress reduction function based on soil–plant water relations. Irrigation Science, 39, 101-122.

Yu, R., & Kurtural, S. K. (2020). Proximal sensing of soil electrical conductivity provides a link to soil-plant water relationships and supports the identification of plant water status zones in vineyards. Frontiers in Plant Science, 11, 512647.

24

Soil Plant Interaction and Nutrient Uptake

Rajni Yadav[1], Anil Kumar[2], Pankaj[3], Dharam Pal[4]

[1,2,4]Department of Soil Science, Chaudhary Charan Singh Haryana Agricultural University, Hisar, Haryana
[3]Assistant Professor, Guru Kashi University, Talwandi Sabo Bathinda, Punjab

Abstract

The interaction between soil and plants is a complex and dynamic process that plays a vital role in determining plant growth, productivity, and nutrient uptake. This abstract explores the intricate relationship between soil properties, root morphology, and nutrient availability, highlighting key mechanisms involved in nutrient uptake by plants. Soil serves as the primary reservoir of nutrients essential for plant growth, providing essential elements such as nitrogen, phosphorus, potassium, and micronutrients. Plant roots play a critical role in accessing these nutrients, with their morphology and distribution influenced by soil texture, structure, and nutrient availability. Various mechanisms, including root exudation, mycorrhiza associations, and ion exchange, facilitate nutrient uptake by plants, enhancing their resilience to environmental stresses and optimizing nutrient use efficiency. Understanding the dynamics of soil-plant interaction and nutrient uptake is essential for sustainable agriculture, guiding the development of management practices that promote soil fertility, crop nutrition, and ecosystem resilience.

Keywords: *soil, nutrient, stress, plants, elements*

Introduction

The intricate dance between soil and plants is a cornerstone of terrestrial ecosystems, underpinning the vitality of agricultural systems and natural habitats alike. At the heart of this symbiotic relationship lies the exchange of nutrients, where soil acts as a reservoir and plants as beneficiaries. This article delves deep into the mechanisms and dynamics of soil-plant interaction and nutrient uptake, exploring the interplay of biological, chemical, and physical processes that shape the flow of essential elements from soil to plant roots.

Understanding Soil-Plant Interaction

The relationship between soil and plants is fundamental to terrestrial ecosystems, influencing plant growth, nutrient cycling, and ecosystem functioning. Understanding this interaction is crucial for sustainable agriculture, ecosystem management, and environmental conservation. This article explores the dynamics of soil-plant interaction, highlighting key processes, factors, and implications.

1. Soil as a Growth Medium

Physical Support: Soil provides anchorage and support for plant roots, allowing them to penetrate the soil profile and anchor the plant securely.

Nutrient Source: Soil acts as a reservoir of essential nutrients required for plant growth, providing elements such as nitrogen, phosphorus, potassium, and micronutrients.

Water Supply: Soil stores and supplies water to plant roots, ensuring adequate hydration for metabolic processes such as photosynthesis and transpiration.

2. Root-Soil Interface

Root Morphology: Plant roots exhibit diverse morphological adaptations to optimize nutrient uptake and water absorption. These adaptations include root hairs, root branching, and root elongation.

Root Exudates: Plants release organic compounds known as root exudates into the soil, influencing soil microbial communities, nutrient availability, and soil structure.

Rhizosphere: The zone of soil surrounding plant roots, known as the rhizosphere, is enriched with microbial activity and nutrient exchange, facilitating nutrient uptake by roots.

3. Nutrient Uptake Mechanisms

Passive Uptake: Nutrients in soil solution are absorbed by plant roots passively, driven by concentration gradients and root membrane permeability.

Active Uptake: Plants actively absorb nutrients against concentration gradients using specialized transport proteins, such as ion channels and carriers, located on root cell membranes.

Mycorrhizal Symbiosis: Mycorrhizal fungi form mutualistic associations with plant roots, enhancing nutrient uptake by extending the root system and increasing nutrient absorption surface area.

4. Factors Influencing Soil-Plant Interaction

Soil Properties: Soil texture, structure, pH, organic matter content, and nutrient levels influence plant growth and nutrient availability.

Environmental Factors: Light intensity, temperature, humidity, and soil moisture affect plant physiological processes and nutrient uptake efficiency.

Plant Characteristics: Plant species, genotype, age, and health impact root morphology, exudation patterns, and nutrient uptake strategies.

5. Implications for Agriculture and Environment

Crop Productivity: Understanding soil-plant interaction enables farmers to optimize soil fertility, nutrient management, and irrigation practices, enhancing crop yields and quality.

Soil Health: Sustainable soil-plant interaction promotes soil health, biodiversity, and resilience to environmental stressors, reducing erosion, nutrient runoff, and soil degradation.

Ecosystem Stability: Healthy soil-plant interactions contribute to ecosystem stability, supporting diverse plant communities, carbon sequestration, and water infiltration.

Soil Factors Influencing Nutrient Availability

Soil serves as a reservoir of essential nutrients required for plant growth, and various soil factors profoundly influence the availability of these nutrients to plants. Understanding how soil properties impact nutrient availability is crucial for effective soil management and sustainable agriculture. This article explores the key soil factors that influence nutrient availability and their implications for plant growth.

1. Soil Texture

Definition: Soil texture refers to the relative proportions of sand, silt, and clay particles in the soil. It greatly affects nutrient availability due to its influence on water retention, aeration, and cation exchange capacity (CEC).

Impact on Nutrient Availability

- **Sand**: Coarse-textured soils have large pore spaces and low water and nutrient retention capacity. Nutrients can leach quickly through sandy soils, leading to lower nutrient availability for plants.

- **Silt**: Medium-textured soils have moderate water and nutrient retention capacity. They provide better nutrient availability compared to sandy soils but may still experience leaching under excessive rainfall.
- **Clay**: Fine-textured soils have high water and nutrient retention capacity but may suffer from poor drainage and aeration. Clay soils can retain nutrients for longer periods, making them more available to plants but may also become compacted, limiting root growth.

2. Soil pH

Definition: Soil pH measures the acidity or alkalinity of the soil, affecting nutrient solubility, microbial activity, and plant root function.

Impact on Nutrient Availability

- **Acidic Soils (pH < 7)**: Acidic soils tend to have higher concentrations of aluminium and manganese, which can be toxic to plants. They also reduce the availability of essential nutrients such as phosphorus, calcium, and magnesium.
- **Alkaline Soils (pH > 7)**: Alkaline soils may limit the availability of micronutrients such as iron, zinc, and manganese, leading to nutrient deficiencies in plants. They also affect the solubility of phosphorus, reducing its availability to plants.

3. Soil Organic Matter

Definition: Soil organic matter consists of decomposed plant and animal residues, microorganisms, and humus. It plays a crucial role in nutrient cycling, soil structure, and water retention.

Impact on Nutrient Availability

- **Nutrient Storage**: Soil organic matter acts as a reservoir of nutrients, releasing them slowly over time as it decomposes. Organic matter decomposition releases nutrients such as nitrogen, phosphorus, and sulphur, making them available to plants.
- **Cation Exchange Capacity (CEC)**: Organic matter increases the soil's CEC, allowing it to hold onto cations such as calcium, magnesium, potassium, and ammonium. This enhances nutrient availability to plants and reduces nutrient leaching.

4. Soil Moisture

Definition: Soil moisture refers to the amount of water present in the soil. It influences nutrient availability by affecting microbial activity, nutrient transport, and plant root uptake.

Impact on Nutrient Availability

- **Water Transport**: Adequate soil moisture is necessary for the movement of nutrients in the soil solution and their uptake by plant roots. Waterlogged soils may restrict oxygen availability and inhibit nutrient uptake by roots.
- **Microbial Activity**: Soil moisture affects microbial activity, which plays a vital role in nutrient cycling and decomposition. Moist soils support microbial activity, promoting nutrient mineralization and availability to plants.

5. Soil Aeration

Definition: Soil aeration refers to the exchange of gases (oxygen and carbon dioxide) between the soil and the atmosphere. It influences nutrient availability by affecting root respiration, microbial activity, and soil redox potential.

Impact on Nutrient Availability

- **Root Respiration**: Adequate soil aeration is essential for root respiration, which provides energy for nutrient uptake processes. Poorly aerated soils may limit root growth and nutrient uptake by plants.
- **Microbial Activity**: Soil aeration affects microbial activity, with aerobic microbes thriving in well-aerated soils and anaerobic microbes dominating in poorly aerated soils. Aerobic microbes play a crucial role in nutrient cycling and organic matter decomposition, releasing nutrients for plant uptake.

Mechanisms of Nutrient Uptake

Nutrient uptake in plants is a complex process governed by various physiological, biochemical, and molecular mechanisms. Understanding these mechanisms is essential for optimizing nutrient management in agriculture and enhancing crop productivity. This article explores the key processes involved in nutrient uptake by plant roots and their implications for plant growth and development.

1. Passive Uptake

Definition: Passive uptake involves the movement of nutrients into plant roots along a concentration gradient, driven by differences in nutrient concentrations between the soil solution and the root cell cytoplasm.

Mechanism

- **Diffusion**: Nutrients dissolved in the soil solution move from areas of higher concentration in the soil to areas of lower concentration at the root surface, where they are passively absorbed by the root cells.
- **Facilitated Diffusion**: Some nutrients, such as urea and amino acids, may require specialized membrane transport proteins to facilitate their diffusion across the root cell membrane.

Implications: Passive uptake is the primary mode of nutrient absorption for ions present in low concentrations in the soil solution. It is influenced by factors such as nutrient concentration, soil moisture, and root surface area.

2. Active Uptake

Definition: Active uptake involves the energy-dependent transport of nutrients into plant roots against a concentration gradient, requiring the expenditure of metabolic energy in the form of adenosine triphosphate (ATP).

Mechanism

- **Active Transport**: Specialized membrane transport proteins, including pumps and carriers, actively transport specific nutrients across the root cell membrane, often against a concentration gradient.
- **Proton Pumping**: In many cases, active transport is coupled with proton pumping, where protons (H^+ions) are actively pumped out of the root cells, creating an electrochemical gradient that drives nutrient uptake.

Implications: Active uptake is essential for the uptake of essential nutrients present in low concentrations in the soil or when the root cell cytoplasm already contains high concentrations of the nutrient. It allows plants to maintain nutrient homeostasis and adapt to changing soil nutrient conditions.

3. Mycorrhizal Symbiosis

Definition: Mycorrhizal symbiosis is a mutually beneficial association between plant roots and certain fungi, known as mycorrhizae, which enhances nutrient uptake and promotes plant growth.

Mechanism

- **Hyphal Exploration**: Mycorrhizal fungi extend their hyphal network into the soil, exploring a larger soil volume than plant roots alone and accessing nutrients in soil microsites inaccessible to roots.
- **Nutrient Transfer**: Mycorrhizal fungi absorb nutrients such as phosphorus and micronutrients from the soil and transfer them to the plant roots in exchange for carbon compounds provided by the plant.

Implications: Mycorrhizal symbiosis enhances nutrient uptake efficiency, particularly for phosphorus, which is often limited in soil. It improves plant resilience to nutrient deficiencies and environmental stresses.

4. Root Exudation

Definition: Root exudation involves the release of organic compounds, such as organic acids, sugars, amino acids, and enzymes, from plant roots into the rhizosphere, the soil region surrounding the roots.

Mechanism

- **Chemical Signalling**: Root exudates serve as chemical signals that influence soil microbial communities, stimulate microbial activity, and promote nutrient mineralization and availability.
- **Nutrient Mobilization**: Organic acids released by plant roots can chelate metal ions in the soil, making them more soluble and available for plant uptake. Sugars and amino acids can also stimulate microbial activity, enhancing nutrient cycling.

Implications: Root exudation enhances nutrient availability in the rhizosphere, promoting plant growth and nutrient uptake. It plays a crucial role in rhizosphere ecology and soil fertility.

Factors Influencing Nutrient Uptake Efficiency

Nutrient uptake efficiency in plants is influenced by a multitude of factors, ranging from soil conditions and plant characteristics to environmental factors and management practices. Understanding these factors is crucial for optimizing nutrient uptake and maximizing crop productivity. This article explores the key factors that influence nutrient uptake efficiency in plants and their implications for agricultural management.

1. Soil Factors

Nutrient Availability: The availability of nutrients in the soil directly affects their uptake by plant roots. Factors such as soil pH, soil organic matter content, cation exchange capacity (CEC), and soil texture influence nutrient availability and uptake efficiency.

Soil Moisture: Adequate soil moisture is essential for nutrient uptake by plant roots. Waterlogged or drought-stressed soils can inhibit root growth and nutrient absorption, reducing nutrient uptake efficiency.

Soil Aeration: Soil aeration affects root respiration and nutrient uptake. Poorly aerated soils may limit oxygen availability to roots, hindering nutrient uptake efficiency.

2. Plant Factors

Root Morphology: Root morphology, including root length, surface area, and branching pattern, influences nutrient uptake efficiency. Plants with extensive root systems can access a larger volume of soil and acquire nutrients more efficiently.

Root Exudation: Root exudates play a crucial role in nutrient acquisition by plants. Plants release organic compounds into the rhizosphere that promote microbial activity and nutrient mobilization, enhancing nutrient uptake efficiency.

Genetic Traits: Plant species and cultivars exhibit genetic variability in nutrient uptake efficiency. Breeding for traits such as root architecture, nutrient transporter proteins, and nutrient use efficiency can improve nutrient uptake efficiency in crops.

3. Environmental Factors

Temperature: Temperature influences plant physiological processes, including nutrient uptake. Optimal temperatures promote root growth and metabolic activity, enhancing nutrient uptake efficiency.

Light Intensity: Light availability affects photosynthesis and root exudation, indirectly influencing nutrient uptake efficiency. Plants grown under low light conditions may exhibit reduced nutrient uptake rates.

Humidity: Humidity levels impact transpiration rates and water uptake by plant roots, which in turn affect nutrient uptake efficiency. High humidity may reduce transpiration and nutrient transport in plants.

4. Management Practices

Fertilization: Proper fertilization practices can enhance nutrient availability and uptake efficiency. Applying balanced fertilizers tailored to crop nutrient requirements and soil conditions can optimize nutrient uptake and minimize nutrient losses.

Irrigation: Efficient irrigation management ensures adequate soil moisture levels for optimal nutrient uptake. Irrigation scheduling based on crop water requirements and soil moisture monitoring can improve nutrient uptake efficiency.

Crop Rotation and Intercropping: Crop rotation and intercropping can enhance nutrient cycling and availability in the soil, promoting nutrient uptake efficiency in subsequent crops.

Implications for Agriculture

Understanding the factors influencing nutrient uptake efficiency is essential for sustainable agricultural management. By optimizing soil conditions, selecting appropriate crop varieties, and implementing efficient management practices, farmers can improve nutrient uptake efficiency, reduce nutrient losses, and enhance crop productivity. Additionally, research into novel technologies and breeding strategies aimed at improving nutrient uptake efficiency in crops can contribute to global food security and environmental sustainability.

Conclusion

The soil-plant interaction and nutrient uptake represent a multifaceted interplay of biological, chemical, and physical processes essential for plant growth, agricultural productivity, and ecosystem health. By unravelling the complexities of this relationship, we gain insights into how soil properties, root morphology, microbial symbiosis, and environmental factors shape nutrient availability and uptake by plants. Armed with this knowledge, we can develop innovative nutrient management strategies that optimize nutrient use efficiency, minimize environmental impacts, and sustainably meet the nutritional needs of a growing global population.

References

Agegnehu, G., Nelson, P. N., & Bird, M. I. (2016). Crop yield, plant nutrient uptake and soil physicochemical properties under organic soil amendments and nitrogen fertilization on Nitisols. Soil and Tillage Research, 160, 1-13.

Dunbabin, V. M., Diggle, A. J., Rengel, Z., & van Hugten, R. (2002). Modelling the interactions between water and nutrient uptake and root growth. Plant and Soil, 239, 19-38.

Fageria, V. D. (2001). Nutrient interactions in crop plants. Journal of plant nutrition, 24(8), 1269-1290.

Loh, S. K., Cheong, K. Y., & Salimon, J. (2017). Surface-active physicochemical characteristics of spent bleaching earth on soil-plant interaction and water-nutrient uptake: A review. Applied Clay Science, 140, 59-65.

Vega, N. W. O. (2007). A review on beneficial effects of rhizosphere bacteria on soil nutrient availability and plant nutrient uptake. Revista Facultad Nacional de Agronomía-Medellín, 60(1), 3621-3643.

25

Soil Amendments and Organic Matter

Rajni Yadav, Anil Kumar, Sekhar Kumar and Khatera Qane

Department of Soil Science, Chaudhary Charan Singh Haryana Agricultural University, Hisar, Haryana

Abstract

Soil amendments and organic matter play critical roles in enhancing soil fertility, improving crop productivity, and promoting sustainable agriculture. This abstract provides an overview of the importance of soil amendments and organic matter in soil management practices, highlighting their impacts on soil structure, nutrient availability, and microbial activity. By incorporating organic amendments such as compost, manure, and green manures into soil, farmers can replenish essential nutrients, improve soil structure, and enhance water retention capacity. Additionally, soil amendments can mitigate soil degradation, erosion, and nutrient runoff, contributing to long-term soil health and ecosystem resilience. Understanding the benefits of soil amendments and organic matter is essential for promoting sustainable soil management practices and ensuring food security in a changing climate.

Keywords: *soil, green, farmers, sustainable, practices*

Introduction

Soil is the lifeblood of agriculture, providing the foundation for plant growth and sustenance. However, over time, intensive agricultural practices, urbanization, and environmental degradation have depleted soil fertility and compromised its health. In response, farmers and land managers have turned to soil amendments and organic matter as powerful tools for rejuvenating soils, enhancing productivity, and fostering sustainability. This article delves into the multifaceted world of soil amendments and organic matter, exploring their significance, mechanisms, types, and implications for agricultural and environmental stewardship.

Understanding Soil Amendments and Organic Matter

The Essence of Soil Amendments

Definition: Soil amendments encompass a diverse array of materials added to soil to improve its physical, chemical, and biological properties. These materials augment soil fertility, structure, water retention capacity, and nutrient availability, facilitating optimal conditions for plant growth.

Types of Soil Amendments

Soil amendments encompass a diverse array of materials added to soil with the aim of improving its physical, chemical, and biological properties. These amendments play a crucial role in enhancing soil fertility, structure, and overall health. Here are some common types of soil amendments:

1. Organic Amendments

- **Compost**: Compost is a nutrient-rich organic material produced through the decomposition of organic waste such as food scraps, yard trimmings, and manure. It improves soil structure, enhances water retention, and provides essential nutrients to plants.

- **Manure**: Animal manure, such as cow, chicken, or horse manure, is a valuable source of organic matter and nutrients. It enriches soil with nitrogen, phosphorus, potassium, and other micronutrients, promoting plant growth and soil fertility.

- **Biochar**: Biochar is a carbon-rich material produced through the pyrolysis of organic biomass under controlled conditions. It enhances soil fertility, improves water retention, and promotes microbial activity, making it a valuable amendment for agricultural soils.

- **Green Manures**: Green manures are cover crops grown specifically to improve soil health and fertility. They add organic matter to the soil, fix nitrogen from the atmosphere, suppress weeds, and prevent soil erosion.

2. Inorganic Amendments

- **Lime**: Lime is commonly used to adjust soil pH and reduce soil acidity. It helps neutralize soil acidity, improve nutrient availability, and enhance soil structure. There are different types of lime, including calcitic lime, dolomitic lime, and hydrated lime.

- **Gypsum**: Gypsum is a calcium sulphate mineral used to improve soil structure in heavy clay soils. It reduces soil compaction, improves water infiltration, and enhances root penetration.

- **Rock Phosphate**: Rock phosphate is a natural mineral fertilizer rich in phosphorus. It provides a slow-release source of phosphorus to plants, promoting root development, flowering, and fruiting.
- **Potassium Sulphate**: Potassium sulphate is a soluble potassium fertilizer used to supplement potassium levels in soils. It improves plant vigour, stress tolerance, and fruit quality.

3. Other Amendments

- **Sulphur**: Sulphur is used to lower soil pH in alkaline soils. It promotes soil acidification, improves nutrient availability, and enhances microbial activity.
- **Vermiculite and Perlite**: Vermiculite and perlite are lightweight minerals used to improve soil aeration and water retention in container gardening and potting mixes.
- **Biosolids**: Biosolids are organic materials derived from sewage sludge. When properly treated and applied, they can serve as a nutrient-rich soil amendment, providing organic matter and essential nutrients to plants.

The Vitality of Organic Matter

Definition: Organic matter comprises decomposed plant and animal residues, microorganisms, and humus, serving as the lifeblood of soil fertility and ecosystem resilience. It fosters soil aggregation, nutrient cycling, water retention, and microbial activity, driving essential soil functions.

Role of Organic Matter

- **Nutrient Cycling**: Organic matter acts as a reservoir of nutrients, releasing them gradually as it decomposes and supporting plant growth.
- **Soil Structure**: Organic matter improves soil structure by enhancing aggregation, friability, and porosity, promoting root penetration, water infiltration, and aeration.
- **Microbial Activity**: Organic matter sustains soil microbial communities, fostering nutrient mineralization, organic matter decomposition, and disease suppression.
- **Water Retention**: Organic matter improves soil water-holding capacity, mitigating drought stress and enhancing plant resilience to water fluctuations.

Mechanisms of Action

Nutrient Cycling and Availability

Organic Matter Decomposition: Organic matter decomposition by soil microbes releases nutrients such as nitrogen, phosphorus, potassium, and micronutrients, making them available for plant uptake.

Mineralization and Immobilization: Soil microbes mineralize organic matter, converting organic nutrients into inorganic forms accessible to plants. Conversely, nutrient immobilization occurs when microbes assimilate nutrients for their growth, temporarily reducing nutrient availability to plants.

Soil Structure and Aggregation

Humus Formation: Organic matter decomposition results in the formation of humus, a stable organic component that binds soil particles together, enhancing soil aggregation and stability.

Porosity and Aeration: Organic matter improves soil porosity and aeration by creating macrospores and enhancing soil structure, facilitating root growth, gas exchange, and water infiltration.

Microbial Activity and Diversity

Microbial Decomposition: Soil microbes degrade organic matter, breaking down complex organic compounds into simpler forms and releasing nutrients for plant uptake.

Rhizosphere Ecology: Organic matter inputs stimulate microbial activity in the rhizosphere, the soil region surrounding plant roots, promoting symbiotic interactions, nutrient cycling, and disease suppression.

Types of Soil Amendments and Their Applications

Soil amendments come in various forms, each with specific properties and benefits that contribute to soil health and fertility. Understanding the types of soil amendments and their applications is crucial for sustainable agriculture. Here are some common types of soil amendments and their respective applications:

1. Organic Amendments

a. Compost

- **Composition**: Compost is a mixture of decomposed organic materials such as kitchen scraps, yard waste, and plant residues.

- **Applications**: It enriches soil with organic matter, improves soil structure, enhances water retention, and provides essential nutrients for plant growth.
- **Benefits**: Compost promotes microbial activity, suppresses diseases, and reduces the need for chemical fertilizers. It is suitable for a wide range of crops and soil types.

b. Manure

- **Composition**: Animal manure, such as cow, horse, or poultry manure, is a rich source of organic matter, nitrogen, phosphorus, and potassium.
- **Applications**: It adds organic nutrients to the soil, improves soil fertility, and enhances microbial activity.
- **Benefits**: Manure increases soil organic matter content, promotes nutrient cycling, and enhances soil structure. It is commonly used in organic farming and sustainable agriculture practices.

c. Green Manures

- **Composition**: Green manures are cover crops grown specifically to improve soil health and fertility.
- **Applications**: They add organic matter to the soil, fix nitrogen from the atmosphere, suppress weeds, and prevent soil erosion.
- **Benefits**: Green manures enhance soil structure, promote nutrient cycling, and provide habitat for beneficial insects. They are often used in crop rotation systems to improve soil quality and fertility.

2. Inorganic Amendments

a. Lime

- **Composition**: Lime is a calcium or calcium-magnesium compound used to adjust soil pH.
- **Applications**: It raises soil pH in acidic soils, improves nutrient availability, and enhances soil structure.
- **Benefits**: Lime reduces soil acidity, increases the availability of nutrients such as phosphorus and calcium, and promotes the growth of beneficial soil microorganisms.

b. Gypsum

- **Composition**: Gypsum is a calcium sulphate mineral used to improve soil structure in heavy clay soils.
- **Applications**: It reduces soil compaction, improves water infiltration, and enhances root penetration.
- **Benefits**: Gypsum improves soil aeration, drainage, and root growth. It is particularly useful in sodic soils with high sodium levels.

c. Rock Phosphate

- **Composition**: Rock phosphate is a natural mineral fertilizer rich in phosphorus.
- **Applications**: It provides a slow-release source of phosphorus to plants, promoting root development, flowering, and fruiting.
- **Benefits**: Rock phosphate enhances soil fertility, promotes healthy plant growth, and improves crop yield and quality.

3. Other Amendments

a. Biochar

- **Composition**: Biochar is a carbon-rich material produced through the pyrolysis of organic biomass.
- **Applications**: It improves soil fertility, water retention, and nutrient availability.
- **Benefits**: Biochar enhances soil carbon sequestration, reduces greenhouse gas emissions, and promotes soil microbial activity and nutrient retention.

b. Sulphur

- **Composition**: Sulphur is used to lower soil pH in alkaline soils.
- **Applications**: It promotes soil acidification, improves nutrient availability, and enhances microbial activity.
- **Benefits**: Sulphur reduces soil pH, increases the availability of nutrients such as iron and manganese, and enhances plant growth and productivity.

c. Biosolids

- **Composition**: Biosolids are organic materials derived from sewage sludge.

- **Applications**: When properly treated and applied, they can serve as a nutrient-rich soil amendment.
- **Benefits**: Biosolids improve soil fertility, promote plant growth, and enhance soil structure and water retention.

Implications for Agriculture and Environment

Sustainable Soil Management

- **Enhanced Productivity**: Soil amendments and organic matter improve soil fertility, structure, and water retention, enhancing crop productivity and yield stability.
- **Climate Resilience**: Organic matter-rich soils exhibit greater resilience to climate variability, mitigating drought stress, erosion, and nutrient leaching.
- **Environmental Protection**: Proper soil management practices, including the use of organic amendments, reduce soil erosion, nutrient runoff, and pollution, safeguarding water quality and ecosystem health.

Conservation and Restoration

- **Soil Conservation**: Soil amendments and organic matter play a vital role in soil conservation efforts, preventing soil degradation, erosion, and loss of arable land.
- **Ecosystem Restoration**: Incorporating organic matter into degraded soils restores soil health, biodiversity, and ecosystem functioning, fostering habitat restoration and ecological resilience.

Challenges and Future Directions

Challenges and future directions in the realm of soil amendments present both hurdles to overcome and opportunities for innovation and improvement. Let's delve into these aspects:

Challenges

1. **Cost and Accessibility**: The affordability and availability of certain soil amendments, especially for small-scale farmers or those in remote areas, pose significant challenges. Finding cost-effective solutions and improving access to quality amendments is crucial for equitable agricultural development.
2. **Quality Control**: Ensuring the consistency and quality of soil

amendments, particularly organic materials like compost and manure, can be difficult. Variations in composition, nutrient content, and potential contaminants necessitate rigorous quality control measures to maintain efficacy and safety.

3. **Environmental Concerns**: Mismanagement of soil amendments can lead to environmental degradation, including nutrient runoff, soil erosion, and water pollution. Balancing the benefits of soil amendments with environmental protection requires careful regulation and adherence to sustainable practices.
4. **Regulatory Compliance**: Compliance with regulations and standards related to the use of soil amendments can be complex and burdensome for farmers. Streamlining regulatory processes and providing support for compliance are essential for promoting responsible soil management practices.
5. **Knowledge Gaps**: Many farmers lack awareness or understanding of the benefits and proper application of soil amendments. Bridging knowledge gaps through education, extension services, and farmer training programs is critical for maximizing the effectiveness of soil amendment strategies.

Future Directions

1. **Innovation in Technology**: Continued research and development of innovative soil amendment technologies, such as biochar production, precision application methods, and microbial inoculants, can lead to more efficient and sustainable soil management practices.
2. **Circular Economy Approaches**: Embracing circular economy principles can help maximize the value of organic materials and waste streams by converting them into beneficial soil amendments. This approach promotes resource efficiency, waste reduction, and sustainable agricultural practices.
3. **Climate-Smart Solutions**: Integrating soil amendments into climate-smart agricultural strategies can help mitigate climate change impacts. Enhancing soil carbon sequestration, reducing greenhouse gas emissions, and promoting climate-resilient cropping systems are essential components of such approaches.
4. **Community Engagement**: Engaging farmers, researchers, policymakers, and other stakeholders in collaborative decision-making

processes fosters knowledge sharing, innovation, and the co-creation of solutions to soil management challenges. Building partnerships and networks strengthens collective efforts towards sustainable soil management.

5. **Adaptive Management Practices**: Implementing adaptive management approaches that incorporate soil monitoring, data analytics, and feedback mechanisms enables farmers to adjust soil amendment strategies in response to changing environmental conditions, crop needs, and market dynamics.

Conclusion

Soil amendments and organic matter represent indispensable tools for sustainable soil management and agricultural resilience in a rapidly changing world. By harnessing the power of organic inputs, farmers can rejuvenate degraded soils, enhance crop productivity, and promote environmental sustainability. However, realizing the full potential of soil amendments requires integrated approaches, stakeholder collaboration, and ongoing research to address challenges and optimize practices for diverse agricultural systems worldwide. As custodians of the land, it is our collective responsibility to nurture and replenish soils for future generations, ensuring food security, ecosystem health, and planetary well-being.

References

Cooperband, L. (2002). Building soil organic matter with organic amendments. Center for Integrated Agricultural Systems (CIAS), College of Agricultural and Life Sciences, University of Wisconsin-Madison.

Goyal, S., Chander, K., Mundra, M. C., & Kapoor, K. K. (1999). Influence of inorganic fertilizers and organic amendments on soil organic matter and soil microbial properties under tropical conditions. Biology and Fertility of Soils, 29, 196-200.

Lima, D. L., Santos, S. M., Scherer, H. W., Schneider, R. J., Duarte, A. C., Santos, E. B., & Esteves, V. I. (2009). Effects of organic and inorganic amendments on soil organic matter properties. Geoderma, 150(1-2), 38-45.

Oldfield, E. E., Wood, S. A., & Bradford, M. A. (2018). Direct effects of soil organic matter on productivity mirror those observed with organic amendments. Plant and soil, 423, 363-373.

Sebastia, J., Labanowski, J., & Lamy, I. (2007). Changes in soil organic matter chemical properties after organic amendments. Chemosphere, 68(7), 1245-1253.

Wei, W., Yan, Y., Cao, J., Christie, P., Zhang, F., & Fan, M. (2016). Effects of combined application of organic amendments and fertilizers on crop yield and soil organic matter: An integrated analysis of long-term experiments. Agriculture, Ecosystems & Environment, 225, 86-92.